Building Global Security Through Cooperation

Building Global Security Through Cooperation

Annals of Pugwash 1989

Edited by
J. Rotblat and J. P. Holdren

Springer-Verlag
Berlin Heidelberg New York London
Paris Tokyo Hong Kong Barcelona

Professor Dr. Joseph Rotblat
Pugwash, Flat A, Museum Mansions,
63A Great Russell Street,
London WC1B 3BJ, United Kingdom

Professor John P. Holdren
Energy & Resources Group, 100 T-4,
University of California,
Berkeley, CA 94720, USA

ISBN-13: 978-3-642-75845-4 e-ISBN-13: 978-3-642-75843-0
DOI: 10.1007/978-3-642-75843-0

Library of Congress Cataloging-in-Publication Data. Building global security through cooperation : annals of Pugwash 1989 / edited by J. Rotblat and J. P. Holdren. p. cm. Based on material presented at the 39th Pugwash Conference, held in Cambridge, Massachusetts, from 23 to 28 July 1989. Includes bibliographical references and index. ISBN 3-540-52813-X (Berlin Heidelberg New York : acid-free paper) – ISBN 0-387-52813-X (New York Berlin Heidelberg : acid-free paper) 1. Nuclear disarmament–Congresses. 2. Chemical warfare (International law)–Congresses. 3. Security, International–Congresses. 4. International cooperation–Congresses. I. Rotblat, Joseph, 1908– . II. Holdren, John P. III. Pugwash Conference on Science and World Affairs (39th : 1989 : Cambridge, Mass.) JX1974.7.B815 1990 327.1'74–dc20
90-10194

Softcover reprint of the hardcover 1st edition 1990

2157/3140-543210 – Printed on acid-free paper

Preface

In the world of international affairs, 1989 was by any standard an extraordinary year. Opening in an atmosphere of old perils and new opportunities in international relations, 1989 closed as an undoubted watershed; the year in which the Cold War as we had known it for 45 years came to an end. Yet while much has changed, much remains unfinished, unresolved, or not changed at all.

Thus, although the sweeping reforms and political restructuring in Eastern Europe have essentially eradicated the ideological and political basis for the Cold War, the bloated military forces of the East-West confrontation – nuclear and chemical as well as conventional – have really only begun to be trimmed. At this writing, in early 1990, there is still no START agreement, no treaty on Conventional Forces in Europe, no formal restraint on nuclear weapons with ranges below 500 kilometres, no prescription for protecting the ABM Treaty against erosion by the mindless pursuit of new capabilities in space weaponry.

Moreover, even as the last vestiges of the rationale for the planned and existing military forces drain out of the East-West relationship, these forces retain tremendous potential to do harm - for example by unintended, perhaps virtually automatic escalation from some modest local instability. We cannot rest, do not dare succumb to complacency, until these arsenals actually have been dismantled. The obstacles in the way of that goal, alas, are still many. Prominent voices on both sides warn against deep and rapid cuts, arguing that the future is unpredictable. Political and economic obstacles materialise, as prescriptions for cuts are seen to entail closing of bases and factories in real communities full of real people. In the Soviet Union, assimilation of demobilised officers and troops back into civilian society is proving to be a major problem.

In the less developed countries of the South, home to three quarters of the world's population, the transformations of 1989 in East-West relations are no watershed at all. The roots of conflict in the South are, first of all, more diverse, more deep-rooted, less tractable than in the North – a volatile mixture of racial and religious hatreds, territorial disputes, frictions of emerging statehood, and the frustrations of poverty, oppression, and manipulation in degrees scarcely imagined by most inhabitants of industrialised nations. And while East-West rapprochment may mean a reduction in competitive manoeuvering, manipulation, and direct intervention in the South by the United States and the Soviet Union, many in the South fear it will also mean reductions in development assistance and, perhaps, diminished restraints on local belligerents.

There are additional reasons for supposing that threats to peace may grow along South-South and North-South axes even as they are dwindling on the East-West one. One is the rapid growth in size and sophistication of military arsenals in the South – a trend that was once largely the work of industrial-country arms exporters who sent weapons south for money or influence. While that still goes on, it is being rapidly supplanted as the main factor by the emergence of indigenous manufacturing capabilities and South-South trade in armoured vehicles, supersonic aircraft, ballistic missiles, and the wherewithal for chemical and nuclear weapons. The major powers in the North have dawdled for decades without managing to achieve either a Chemical Weapons Convention or a Comprehensive Test Ban on nuclear explosions, both of them vitally important for stopping horizontal as well as vertical proliferation of these classes of armaments; the growing interest in such capabilities in the South is making pursuit of these agreements both more urgent and more difficult.

Frustrations in the South arising from economic and environmental deprivation seem certain to grow in the years and decades ahead. The existing burden of debt is a staggering drag on development prospects, and financial and technological development assistance from the North remains a pathetically tiny trickle into an ocean of need. Soils, forests, and groundwater are dwindling, while the numbers of people who must be supported on these resources are growing – 90 million people per year worldwide, 90 per cent of them in the South. Global climate change from accumulating greenhouse gases in the atmosphere will have its severest effects in the South, where people have, at the outset, less to eat, poorer health, worse medical care, and less capacity to adapt. What they will probably experience from global climate change is more dry-season heat and drought; more wet-season floods and storm victims; more pest outbreaks, famine, and disease; and more environmental refugees streaming across international borders. There will be a strong tendency to view the causes of all this as primarily the responsibility of the North, by a southern majority already smoldering with resentment over a widening North-South prosperity gap they also regard as mostly the North's fault. The implications for international tension and conflict are all too clear.

The foregoing considerations suggest that the international security agenda needs to be seen as containing three distinct albeit inter-related classes of problems, or sub-agendas, which one could call the East-West Arms Control Agenda, the North-South Arms-Control Agenda, and the Common-Security Agenda. All three of these sub-agendas have been well represented in the deliberations of the Pugwash Conferences on Science and World Affairs for many years, and they are well represented in this volume, which draws particularly heavily on material presented at the 39th Pugwash Conference, *Building Global Security Through Cooperation,* held in Cambridge, Massachusetts, from 23 to 28 July 1989.

The East-West Arms-Control Agenda focusses mainly on limiting offensive and defensive strategic forces, space weaponry, and nuclear and conventional forces in Europe and at sea, as well as on issues of East-West confidence building, crisis avoidance, and crisis control. It is covered in this volume in Part I,

Strategic Forces and Their Limitation, and in Part II, *European Security.* Although nearly as much remained to accomplish on this sub-agenda at the end of 1989 as at the beginning – if accomplishment is measured by actual agreements – the progress during the year in terms of elimination of obstacles was, quite obviously, unprecedented. Gone, almost overnight, is any semblance of a rationale for the sizes and kinds of forces that have come to be arrayed against each other by East and West. Gone, certainly, is the basis for "flexible response" and "extended deterrence" – that is, for "first use of nuclear weapons if necessary", and for all the deployments intended to make that doctrine plausible – although this change has not yet been officially conceded. Much work will still be needed to bring about the reductions that are implied, but surely the momentum in this direction will now become irresistible.

The problems on the North-South Arms-Control Agenda are more multilateral and more complicated, and the prospects of success are considerably more uncertain. They include problems with strong East-West dimensions, such as chemical weapons and nuclear testing, which, however, are inextricably entwined with developments in the South; and they include problems of arms trade, intervention and non-intervention, and regional crisis management whose axes are *mainly* North-South or South-South. These issues are discussed here in Part III, *Chemical Weapons and Their Control,* Part IV, *Nuclear Weapons Proliferation and Testing,* Part V, *The Arms Trade and Regional Conflict,* Part VI, *Role of United Nations and Multilateral Organizations in International Security,* and in Appendix B, *Statement of the Pugwash Executive Committee on the Fourth NPT Review Conference.* The problems in these categories have tended until now to take a back seat to those on the East-West Arms-Control Agenda described above, but to continue to keep them in this secondary status could be a prescription for disaster. Two of the issues in particular – the need to achieve a Comprehensive Test Ban proscribing nuclear explosions of all yields in all media, and the need to block the transfer to the South of newly surplus weapons from the dwindling East-West confrontation – will provide key tests of the insight and the will of leaders in the North. (So far, the Bush administration has not conceded that a CTB would be in the United States' interest; and no major power in the North has been willing to forgo the monetary and political currency obtained from arms transfers to the South.)

The issues we place on the Common-Security Agenda are those often described under the heading of "non-military aspects of security", including matters of trade, finance and debt; issues in scientific and technological cooperation; and the intersection of problems of energy, environment, and development. The Pugwash Conferences have played an important role in recent years in elucidating the connections between these questions and international security, and in arguing that only a concerted, fully cooperative, international approach offers any hope of preventing the problems in this category from undermining the long-term prospects for prosperity and security worldwide. These issues are treated in this volume in Part VII, *Energy-Environment-Development Interactions,* in Part VIII, *East-West and North-South Cooperation,* and, together with all the other topics, in Appendix A, *Statement of the Pugwash Council on the 39th Pugwash Conference.*

We close this preface, as we began it, with an observation on the significance of the events of 1989. What brought about the transition in East-West relations that became so dramatic and irreversible at the end of 1989, was above all, we think, the recognition that pursuing security through East-West military competition had been a complete failure – that it had provided only declining security and increasing danger, at escalating and finally intolerable economic cost. It became a matter of the sheerest necessity, then, to replace the competitive-security paradigm with a cooperative-security approach, in order to shrink the danger and to free for worthier purposes the resources and energies that the confrontational approach had been consuming. What will be required to address successfully the still larger security agendas of the 1990s and beyond – what we have here called the North-South and Common-Security agendas – is the generalisation of this same recognition to a global scale: the problems of East and West and North and South are intertwined; what threatens one group threatens all; security can only be gained through cooperation against those common threats. To understand this and to act on it are the central challenges of our time. We hope this book will make some contribution toward these ends.

The Editors wish to express their gratitude to Dr. David Carlton for help in the editing of the book, and to Mrs. Sue England for assistance in the preparation of the camera-ready copy.

London, Berkeley
August 1990

Joseph Rotblat
John P. Holdren

Contents

Part I

Strategic Forces and Their Limitation

Prospects for Strategic Offensive and Defensive Forces

Richard Garwin

1. Introduction and Status

This is a brief qualitative discussion of developments in offensive and defensive strategic weaponry, and of prospects for their limitation.

On the Soviet side, strategic offensive forces consist mostly of silo-based ICBMs with multiple, independently targeted re-entry vehicles (MIRVs) including 308 SS-18 missiles with ten warheads each. In addition, Soviet forces contain a few mobile SS-24 missiles on special trains, each with ten warheads, plus a single-warhead truck-mobile SS-25 missile system. Table 1 shows that US and Soviet offensive force composition, as reported in *The Military Balance 1988–1989* [1].

On the strategic defensive side, Soviet weaponry includes the Moscow Anti-Ballistic Missile (ABM) system, permitted under the 1972 ABM Treaty, which incorporates up to 100 nuclear-armed interceptors, plus technology efforts characterized by Soviet leaders as including "everything that the United States is doing". This *presumably* includes ground-based free-electron lasers, which might eventually be used with space-deployed relay mirrors in a high-technology defence against strategic ballistic missiles in their boost phase. Such lasers would at an earlier stage be available for anti-satellite activity. The defensive programme presumably includes also investigations of non-nuclear-armed interceptors, of neutral particle beams, and the like. Yet, little actual evidence is available on the scope and status of the Soviet programme, despite talk of "open lab- oratories". In addition, for the last four years or so, Soviet leaders have promised that they will counter any deployment of US defensive systems by "means that are asymmetric", presumably including space mines, anti-satellite weapons, decoys, single- warhead boosters, fast-burn boosters, and the like.

The Soviets have had for decades a vast operational air- defence systems, deploying some 10 000 ground-based radars, a similar number of surface-to-air missiles (SAMs), and several thousand interceptor aircraft. Substantial development is done in support of this costly system, which has no analogue in the United States since the US air-defence system was disbanded in the 1960s, in recognition of the vulnerability to nuclear-armed ballistic missiles.

As for the longer term, the vision of President Mikhail Gorbachev, presented in January 1986 at Reykjavik, has been moved to the indefinite future – the total elimination of nuclear weapons being replaced by an interest in strong reductions, while retaining strategic stability, that is what might be called "minimum deterrence".

Table 1

Name	*Launchers*	*Warheads*	*Yield (kiloton)*	*Comments*
US Forces (June 1988):				
Minuteman II	450	1	1200	In silos
Minuteman III	211	3 MIRV	170	In silos
Minuteman III	300	3 MIRV	335	In silos
MX	39	10 MIRV	300–400	In silos
Poseidon C-3	256	10 MIRV	40	In 16 SSBN
Poseidon D-4	384	8 MIRV	100	In 20 SSBN
plus 362 bombers – B-52G, B-52H, B-1B.				
Soviet Forces (June 1988)				
SS-11 Mod 2	420	1	1000	In silos
SS-11 Mod 3	total	3 MIRV	100–300	In silos
SS-13 Mod 2	60	1	600	In silos
SS-17 Mod 3	138	4 MIRV	500	In silos
SS-18 Mod 4	308	10 MIRV	500	In silos
SS-19 Mod 3	350	6 MIRV	550	In silos
SS-24 solid fuel	10	10 MIRV	100	Rail based
SS-25 solid fuel	100	1	550	Rd. Mobile
Soviet SLBMs:				
SS-N-5	36	1	1000	In 12 SSB
SS-N-6 Mod 1	256	1	500–1000	In 16 SSB
SS-N-6 Mod 3	total	2 MIRV	500	total
SS-N-8 Mod 1	286	1	500–1000	In 23 SSB
SS-N-8 Mod 2	total	1	800	total
SS-N-17	12	1	500	In 1 SSBN
SS-N-18 Mod 1	224	3 MIRV	500	In 14 SSBN
SS-N-18 Mod 2	total	1	500–1000	total
SS-N-18 Mod 3	total	5 MIRV	500	total
SS-N-20	100	6 MIRV	100	In 5 SSBN
SS-N-23	64	10 MIRV	100	In 4 SSBN

On the US side, the land-based forces are being modernized by the deployment of 100 or fewer MX missiles, of which 50 exist in silos now. There is a programme to move these 50, or perhaps to increase their number to 100, deploying them in garrisons in which they would be vulnerable to even inaccurate nuclear attack, but allowing them to be moved with few-hour strategic warning onto the national rail network, so that they would be invulnerable unless their location could be known with some accuracy.

At the same time, there is substantial support (especially in the US Congress) for a mobile deployment of the single-warhead Midgetman, of which some 500 launchers would be deployed on trucks. This follows the teachings of the Scowcroft Commission reports of 1983 and 1984, which proposed that 100 MX missiles be placed in silos, as admittedly vulnerable but still potentially useful weapons, while the main reliance for the long-term future was on the single-warhead Midgetman in a survivable basing mode, and on small submarines each carrying fewer heads than at present. There has been some recent recognition of the fact that a single-warhead Midgetman in a silo is to a large extent self-protecting, since to destroy it would require at least two warheads, even if highly accurate, in order to obtain the reliability of destruction that is required. Furthermore, silo attack would have to reckon with the prospect of launch of the Intercontinental Ballistic Missile (ICBM) force under attack, and thus would be strongly deterred.

As for the sea-based systems, the Trident-II (D-5) missile is in the later stages of testing, with about the usual number of failures. In addition, nuclear-armed sea-launched cruise missiles (SLCMs) have been deployed on a wide range of naval ships.

The air-breathing component of the US strategic force is being modernized with the operational B-1 bomber, equipped with air-launched cruise missiles (ALCMs) as is the B-52 force, and which presumably has the capability to penetrate Soviet air defences, although the electronic countermeasures equipment (ECM) of the B-1 is admittedly much deficient compared with its goal. The B-2 (Stealth) bomber has been revealed to the public eye, and flew for the first time in July 1989. It is doubtful that it will be produced, given the financial strictures on US programmes. The only mission identified for the B-2 is to search out and destroy "strategic relocatable targets", mostly mobile missiles. It is pretty clear that this could not be done in the environment of nuclear war, especially when the other side will not only attempt to hide its mobile missiles but to provide decoys simulating them, and could launch them before they were destroyed if there were a real threat of their destruction. Unfortunately, the US Air Force has recently testified that unless Congress approved the B-2 Bomber Programme, it will withdraw its support for the START negotiations.

As for the US defences, the Anti-ballistic Missile system operational from 1975–1976, incorporating 100 nuclear-armed interceptors, (the "Safeguard" system) was destroyed after 1976, although it was in compliance with the 1972 ABM Treaty; it was judged to cost more than the benefit that it provided. In recent years, since 1984, the Army Ballistic Missile Defense Program has been largely subsumed or directed by the Strategic Defense Initiative Organization (SDIO), and there has been a substantial tension between the long-term programme and "maintaining an option for early deployment", a requirement that was added to the SDI mission shortly after the programme was created. To many, maintaining this option involved building a lobby for early deployment, which would, of course, for any meaningful defence, contravene the 1972 ABM Treaty.

Since President Ronald Reagan's speech of 23 March 1983 and the creation of the SDIO in 1984, expectations have continually declined as to the performance of a system based on either directed-energy weapons or space-based interceptors, to the point where one talks about the possible benefits of a defensive system in countering accidental or unauthorized launch of a small number of weapons, or of "complicating the job of "military planners" on the other side. Specifically, knowledgeable people characterize the capability of a full-scale SDI system as being able to destroy perhaps 10-20 per cent of the currently available strategic offensive warheads, and Harold Agnew (Vice-Chairman of the 1983 Fletcher Panel that defined the SDI) writes [2]:

> *"The most optimistic present analysis of SDI deployments with full bells and whistles might at best destroy half of the assumed ICBM threat if it were to be launched against us. Such an effective deployment, if at all possible, would take, in my opinion, over 20 years to implement fully and would cost in excess of a trillion dollars."*
>
> *"Recently, we are led to believe that the Soviets have accepted in principle the President's original suggestion of a 50-per-cent reduction of strategic forces over the next five to ten years if we will adhere to the narrow interpretation of the ABM Treaty for the next ten years and don't field any ABM system beyond that allowed by the treaty during that period. I would urge subscribing to such an agreement.*
>
> *Clearly, if we actually can eliminate half of the Soviet inventory with a signature, rather than through a premature expenditure of a trillion dollars on an active system employing immature technology which may not actually even be that effective, we should do it . . . ."*

Because the SDI programme has not provided an option that is attractive in comparison with the cost of deployment and the cost of abandoning the ABM Treaty and of countering the counter-measures that will be taken by the other side, the effort has not moved beyond research and demonstration. It has been subject to recurring fashions and enthusiasms, usually sufficient to carry the programme through another year of funding. The current enthusiasm is for "brilliant pebbles" (BPs), which are merely smaller size and more intelligent (not "shiny") versions of "smart rocks" – an endearing term for space-based interceptors (SBI). In fact, although extending hyperbole in two dimensions, brilliant pebbles are certainly not small, being characterized in recent articles as about one metre long and weighing 50 kilogrammes. Although their principal proponent, Lowell Wood of the Lawrence Livermore Laboratory, claims that such BPs can be produced and deployed for \$20 000 each, with each one containing the intelligence of a Cray-1 supercomputer, Lt. Gen. George L. Monahan, Director of the SDIO, in an interview with *The New York Times*, says that cost estimates range from \$275 000 to \$1 million each. If the SBIs are useful only against offensive missiles in boost phase, only those SBIs deployed close to the launch-site at the time of launch can have any capability of destroying boosters, and the cost of the system to handle even the present 2000 boosters can be substantial. The relatively few SBIs that could be so involved could be destroyed by Anti-satellite (ASAT) means; they all could be vulnerable to space mines; and the shortening of boost phase by going to solid-fuel boosters, without significant extension of current technology, would render proposed SBIs (including brilliant pebbles) totally ineffective.

On the political side, although there are calls for a "third-strike counterforce" as being essential for real deterrence of nuclear war, there is far more acceptance of the idea that between the United States and the Soviet Union, nuclear weapons cannot be used other than to destroy valuable targets in the other country after nuclear war has started, and thus to deter the initial strike to begin with. A Senate Hearing on 16 May 1989, chaired by Senator Joseph Biden, had Edward Warner, Barry Blechman, and Stephen Larrabee all discussing the question of nuclear weapons in Europe. Senator Biden asked the three witnesses to explain why deterrence required nuclear weapons on the ground in Europe, but the answers really provided no analytical or political explanation. If one believes that American nuclear weapons are under the control of the American President, it would seem that strategic nuclear weapons kept on the national territory and in the deep oceans of the world, as is the case with submarine-launched ballistic missiles (SLBMs), can do everything that can be done by forward-deployed weapons, at substantially lower cost and with better control. To argue otherwise is to suggest in some way that nuclear weapons might be used deniably, and that the receiving side of the nuclear weapons would be motivated to ignore whence they had come.

The prospects for change in the offensive forces are continued improvement in accuracy, together with a possible (undesirable to almost everyone) reduction in the travel time for some of the nuclear weapons, as in depressed-trajectory launch of SLBMs.

2. What Does This All Mean?

2.1 Deep Cuts

It was probably unrealistic a year ago to have expected action on what would certainly be a major first step towards radical reductions in numbers of nuclear weapons – concluding a Strategic Arms Reduction Talks (START) Treaty by Reagan and Gorbachev at the Moscow Summit in June 1988. Even with the recognition in Washington on 23 March 1988 that there would have had to be a parallel agreement on "space and defence" at the same time as the START agreement, the two sides needed to pull a pair of very large rabbits out of the hat in the two months remaining for negotiation, agreement, and review of the Treaty documents.

In a sense, the first step (a reduction in strategic nuclear weapons by a factor of two) is equivalent to all remaining steps towards the abolition of nuclear weapons. And so it is, but in *number* of nuclear weapons only. No person of common sense will believe that the elimination of the first half of nuclear weapons is as important as the elimination of the second half, and this common-sense position is justified by the concept of "economic targeting" of nuclear weaponry. I have long used this term to indicate not that nuclear weapons are targeted against "economic targets" (those valuable to the "economy") but simply that when nuclear weapons are used, they will be applied against those targets in a given class that will require the fewest nuclear weapons for destruction of a given "value" of target, where value is determined by the opinion of the attacker. Since the first 6000 targets on a list ordered by value have more total value than the second 6000 targets (and vastly more in almost any target population), the overall destructive capability of the strategic forces will not be much reduced in this first step. Additional warheads (in economic targeting) are assigned to additional low-value targets, or are used to place a second or third warhead on a given target to make up for the lack of perfect reliability of the first warhead assigned to that target. Indeed, in the 1983–1984 initial stages of the SDI programme, there was much talk of "devaluing nuclear weapons", and thus of "reducing the incentive for the Soviet Union to build additional nuclear weapons". However, given the fact of "economic targeting", the introduction of a defence that randomly eliminates 90 per cent of the attacking warheads can, in principle, *increase* the value of 10 marginal warheads added to the attacking force (even though only one will penetrate to its target, on the average) in comparison with 10 marginal warheads added to an unopposed attack- ing force - which are anticipated surely to destroy their targets of negligible value! The practiced bureaucrat or proponent, of course, complains that any reduction in his or her programme will eliminate the *most important* weapons and not the *least important* weapons; this is the difference between argument and reason.

2.2 Do We Want Racical Reductions in Nuclear Weaponry?

Many believe that the 50 000 nuclear weapons in the world now are vastly excessive, and that they should be reduced – even to zero. In countries that have very large numbers of nuclear weapons, some ordinary people as well as elites believe that a large number of nuclear weapons are good, and perhaps more is even better. Certainly, the fact that these weapons were acquired at great cost is no reason, in itself, either to retain them or to reduce them.

Retaining things that are costly to buy, costly to maintain, and may constitute a terrible hazard, does not make sense unless there is some *value* associated with their retention. Alternatively, eliminating something of great cost is of no particular benefit unless some of that cost can be recovered. Even with no nuclear weapons in the possession of a potential adversary, there are limits on the number of nuclear weapons that would be acquired by a nation. Each has some cost, and with "economic targeting", (which means sensible plans for use – if they are to be used), there will be a point when the value of a potential target, multiplied by the probability of destroying it if the weapon is used, multiplied by the probability of use of weapons, becomes less than the cost of the weapon itself. But there are other comparisons to be made as well.

When the other side has nuclear weapons there is an inherent question of *pride*, which should not be taken over without question to the nuclear era. However, a number of nuclear weapons entirely adequate facing a non-nuclear adversary might be inadequate when they are subject to destruction before launch, or to interdiction before striking their targets. If one achieved by achievement of design the ability to have one's nuclear weapons survive and to penetrate to their targets, what is the driving force to reduce them? In part it is cost; in part, however, it is to reduce the threat to one's own nation and allies, by encouraging the reduction of weapons of the other side. When one brings in the consideration of third nations (and China, France, and Great Britain are three of the "third nations"), there are arguments both to retain and to reduce. It is difficult to persuade those with a nuclear surfeit that they ought to eliminate so many of their weapons that they would have just about the same number as those nations to which they had been vastly "superior". Of course, it cannot be argued that any particular nation has a "right" to nuclear weapons, unless all nations have a right to nuclear weapons. Nevertheless, nuclear weapons exist, and it is generally believed in the world (and, I believe, correctly) that world security would not be served by having vast numbers of nuclear nations. I conclude that the nuclear giants will not reduce to the level of Great Britain, France, and China. I further conclude that the giants will not march very far down the road of radical reductions unless there is some predictability to the arsenals of these other nations. In particular, if any of these third nations intends to build its weapons to a level of 1000 or 2000, it is unlikely that the superpowers will reduce much below the 6000 strategic nuclear warheads in the first step of START.

In 1988, both Ian Stewart of the British Ministry of Defence, and Sir Geoffrey Howe, the British Foreign Secretary, stated that 50 per cent reductions in

START would not be nearly enough to make the British rethink their plans for modernization of their nuclear forces. And Andre Giraud, then French Minister of Defence, appeared to take the position that START and even further reductions have nothing at all to do with the modernization plans of France. I simply observe that while these nations cannot themselves bring about the radical reductions in US and Soviet forces that they and many others hope for, they can certainly *prevent* such reductions if they do not provide some confidence that their own forces will remain substantially less than the declining forces of the superpowers. Shared security arrangements might reduce these disparities, but that is for the future.

In my opinion, there are two major benefits of radical reductions. The first is to damp the arms race between the United States and the Soviet Union, and the second is to provide a basis for strict opposition to the acquisition of nuclear weapons by any other nations. To this end, probably also a total ban on nuclear tests will eventually be required.

2.3 Discriminate Deterrence?

A 1988 report, funded by $1.6 million of US Government money, proposed a future emphasis for the United States on non-nuclear *strategic* weapons [3]. The commission preparing the report was headed by Fred Ikle and Albert Wohlstetter, and included as members both Z. Brzezinsky and Henry Kissinger. A key assertion in the report is that a decade ago neither US nor Soviet missiles were accurate enough to destroy very hard silos with nuclear weapons, and a decade hence both nations will have missiles so accurate that they could destroy hard silos with non-nuclear strategic weapons. Although the commission notes that this vulnerability would not apply to targets that are underground, mobile, or hidden, it is not clear that they understand how effective these techniques are for survival against non-nuclear strategic weapons. In addition, there is the clear possibility of *defence* against such weapons.

The commission clearly identifies such non-nuclear strategic weapons as missile-delivered. Long, thin rods or cones can penetrate the atmosphere, but it would be very difficult to give them the few-foot accuracy required to destroy a silo or other small strategic target. Nevertheless, homing systems or designators might be developed for this purpose. However, if the targets are under tents, or they are on buses indistinguishable from those that carry civilians, it will not be possible to attack them. Furthermore, even visible hardened targets such as silos can be *defended* readily against long-rod penetrators by a large-scale analogue of the explosive-backed armour now in wide use on modern tanks. A rod which re-enters at hypersonic speed would penetrate several times its length into the silo cover, but if it first had to traverse a 5-centimetre-thick steel plate mounted at an angle of 45 degrees and backed by an equal amount of explosive, the rod would continuously be sheared by the moving steel and would lose most of its penetrating capability.

Similarly, the total *elimination* of strategic weapons would open the door to threats from a few clandestine weapons, from chemical or biological weapons, or even from these non-nuclear weapons. It is for that reason that *nuclear* deterrence will be relied on for a long time, by retaining a certain number of strategic weapons far smaller than the present number. It is to be noted that strategic nuclear weapons could also be used to provide *battlefield* nuclear explosions, and would thus serve the function of reducing the capability of armies by posing the *threat* of massive destruction if they concentrated to gain maximum advantage as in a non-nuclear conflict.

2.4 Elimination of Sea-Based Nuclear Weapons

US nuclear weapons based at sea do not have the permissive action links (PAL) that make it impossible for other US nuclear weapons to provide a nuclear explosion without the communication of the secret word for that particular weapon. Furthermore, the outbreak of nuclear war at sea is certainly not in the interest of the US Navy, and the carriage of nuclear weapons has many important cost elements, ranging from personnel specialized in nuclear weaponry, to special security requirements, to political costs with respect to allies and neutrals. In the pre-ICBM era, there may have been some military and technical justification for nuclear weapons on board ships, but, in fact, the US Navy has recently for strictly military and cost reasons begun to abandon nuclear-armed torpedos, nuclear-armed SAMs, and essentially all nuclear weapons except cruise missiles, air-delivered weapons, and SLBMs. The elimination of all nuclear weapons from sea-based platforms (except SLBMs in specialized submarines) would benefit all.

2.5 Destruction of Nuclear Warheads

The Intermediate Nuclear Forces (INF) Treaty broke new ground in verification, and it laid the *basis* for the destruction of nuclear warheads, although not a single nuclear warhead is required to be destroyed under the INF Treaty. Rather, the heat shield is being removed and destroyed, but the guidance systems and the nuclear explosive itself can be retained and even used on another delivery vehicle.

Frank Carlucci, who was then US Secretary of Defense, explained in testimony supporting the INF Treaty that the United States had not proposed the destruction of the nuclear warheads themselves, but he went on to say that he did not know *how* one would destroy a nuclear warhead – implying that it would be expensive, dangerous, and something that we did not know now how to do. Experts from the US Department of Energy commented to the press, however, that we destroy nuclear warheads every day, in that they come back as obsolete nuclear weapons, are disassembled, the high explosive detonated or burned, the tritium – used for boosting the fission process – removed and reserved for future use, and the plutonium and U-235 reprocessed for further use. Indeed, for demilitarization of nuclear warheads, all these things would be done. The guidance system would be crushed, the explosive removed and destroyed; the tritium removed and retained to replace tritium lost by radioactive decay. However, the plutonium and U-235 would be deformed by being sawed, pressed, or otherwise made impossible for immediate use in a nuclear explosive. These fissile materials would go to safeguarded inventory under the International Atomic Energy Agency (IAEA) in the host country.

In order to make the fissile materials unavailable for a much longer time, the U-235 could be diluted with normal uranium from the 95 per cent most useful in nuclear explosives to a level of about 3 per cent most useful in nuclear reactors. The amount of enrichment effort and time required to bring this low-enriched uranium back to weapon-grade is comparable with that required to start from natural uranium. To make the plutonium less useful, one can mix it with fission products and then store the plutonium under IAEA safeguards. One could imagine, also, adding the possibility of conventional explosives being maintained by the other side in the plutonium storage site, in order to ensure that the safeguards regime is not abrogated – with full knowledge that the use of these explosives would result in their use in one's own safeguarded inventory. Contrary to assertions by some, the actual cost to "de-enrich" U-235 (by dilution with normal uranium) is very small.

2.6 Space Weapons

The question of space weapons has plagued the arms control process. In the joint statement at the end of the Washington Summit on 10 December 1987, Reagan and Gorbachev

> "... *instructed their delegations in Geneva to work out an agreement that would commit the sides to observe the ABM Treaty, as signed in 1972, while conducting their research, development, and testing as required, which are permitted by the ABM Treaty, and not to withdraw from the ABM Treaty for a specified period of time* ... ."

Apparently, this text concealed a major disagreement revealed by the comma preceding the clause "which are permitted by the ABM Treaty". Without the comma, the meaning of the statement would be quite different, with the "which are permitted by the ABM Treaty" limiting the conduct of the research, development, and testing. In any case, commas do not translate very well, and the appearance of agreement in that communique could serve only for a few months instead of a real discussion of the problem. In fact, on 23 March 1988, then-Secretary of State George Shultz and Foreign Minister Edouard Shevardnadze agreed that a separate "defence and space" treaty must be signed together with the START agreement. Commenting on this, the former Director of the Arms Control and Disarmament Agency (ACDA) Kenneth Adelman observed that for three years the US Administration had had the goal essentially of agreeing on *nothing* with the Soviets as regards defence against strategic ballistic missiles and space weapons.

Yet, much of the testing to determine whether SDI is *feasible* can be done totally within the limits of the ABM Treaty. It would only be required to test in support of research and development for the allowed fixed site. Weapons that might nominally be based in space could instead be launched on a rocket for the occasion, and other experiments could be done without impinging in any way on the ABM Treaty limitations. SDIO Director Monahan has now testified that SDI tests can remain within the ABM Treaty limits for the next three or four years.

Another innovation in the space weapons field arose from a speech by Senator Sam Nunn [4]. Expressing profound scepticism about the utility of SDI, he nevertheless proposed two separate but compatible goals for a redirected strategic defensive effort:

> "*...For the near term, we should seriously explore the development of a limited system for protecting against accidental and unauthorized missile launches.*"
>
> "*...For the longer term, our goal should be to pursue research on advanced defensive technologies, principally in the directed energy area, that offer the best prospects for a possible comprehensive defense.*"

The press gave the strong impression that Nunn proposed the deployment (or at least development) of this Accidental Launch Protection System (ALPS), but it is clear from his speech that he did not. He set up a challenge for the SDIO to design such a system which "might be possible within the terms of the ABM Treaty or, at most, require a modest amendment". But Nunn insists, "Such a limited defense would of course have to be proved both technically feasible and affordable. We would also have to carefully consider the extent to which the other elements of the (Paul) Nitze criteria would apply". So, Senator Nunn is proposing that SDIO see whether they can design such a system, but it will not be developed (and certainly not be deployed) unless it is judged feasible, affordable, and (presumably) better than alternative approaches to achieving the same goal.

Unfortunately for SDI (but fortunately for those who fear accidental or unauthorized launch), a solution to that problem is available sooner and at lower cost than via a system of 100 ground-based rockets carrying non-nuclear exo-atmospheric interceptors, and assigned to their targets by powerful radars. The key to the low cost and effectiveness of a proper ALPS system is that the other side want to *cooperate* in eliminating or countering its *own* accidental or unauthorized launch. Thus, if each side retains on its operational missiles those same command-destruct links used in missile flight tests, missiles launched by accident or without authorization could be destroyed by radio command. Furthermore, in addition to broadcasting the command to destroy the missile in flight, the side whose missiles have been launched by accident or without authorization could use the "hot line" to send the particular secret words for the missiles that had been launched, to enable the other side also to broadcast and so to destroy them before they reached full speed and before the warheads could be armed. No nuclear explosions would result, and the likely effectiveness of such a system is far greater than one dependent on missiles to counter other missiles in flight. In fact, the command-destruct system could probably be available in two years and for less than $1 billion. It is hard to argue that even if one wants an Extra-atmospheric Intercept System (ERIS) to *add* security against accidental launch, that one should not obtain that same security sooner by means of the command-destruct link.

Sometimes I am asked "Suppose the other side gets the entire list of command-destruct words? Would that not eliminate the deterrent totally?" Yes, but it would be a poorly-run security system that would allow such an achievement. In fact, one could imagine each of the command-destruct words as being two separate words, held in two locations, and only searched for and produced when it is necessary to destroy that particular missile in flight. There are far greater vulnerabilities to espionage than this one! Finally, it could be arranged that authorized launch could also *disable* the command destruct box before launch, eliminating the possibility of command destruct.

The SDIO-proposed system is expected to be the deployment of 100 ERIS missiles on a single base near Grand Forks, North Dakota, working with the infra-red early-warning satellites and with the various early-warning radars on the periphery of the United States, and also in Great Britain and Greenland. A very interesting analysis by T. A. Postol of Stanford University, however, showed that in order to protect coastal cities against SLBM launch, the ERIS will have to be fired before any radar can see the SLBM in boost phase, and he noted that in recent years there have been thousands of "false alarms" that might have been missile launchings detected by any early-warning satellite.

Worse, to my mind, is the conclusion that a near-Treaty compliant deployment of 100 ERIS would surely have no capability against the Soviet force that would exist at the time of the ALPS deployment. A chart presented by the Lockheed Corporation supporting their proposal for such a development and deployment suggests that a 100-ERIS system could handle 44 objects midcourse. Since the SS-18 Soviet missile has 10 warheads, this might be five SS-18s. But the deployment of ALPS (if it were at all believable for protection against accidental and unauthorized launch) would surely be very good against a single or a few warheads directed in "flexible response" against a few military targets. Thus, the Soviet Union would be certain to deploy large numbers of exo-atmospheric decoys with their warheads. Postol showed how a single SS-18 could deploy hundreds of credible decoys, each one designed to fool the long-wave infra-red homing device of an ERIS interceptor by heating resistive elements on the inside of an inflated plastic balloon. According to Postol, five ounces of lithium battery would suffice to mimic the heat radiated from an ordinary SS-18 warhead during the critical period, and the requirements would be far less if the Soviets instead reduced the infrared from their warhead by improved thermal insulation or even a low- temperature shield to intercept the heat.

Senator Nunn quoted:

"We are increasingly a society of networks – electricity, grids, water systems, oil and gas pipelines, telecommunications links – with highly vulnerable nodes to which we have given virtually no thought of protection, even against the simplest efforts to knock them out. One need look no further than Chernobyl, Bhopal and the recent oil spill in Pittsburgh to appreciate the reality of modern society's vulnerability to catastrophic disruption."

"If we have a finite amount of money to spend and want to spend it in the wisest ways to protect ourselves, our children and our grandchildren, we must seriously assess whether voting a very large share of it to deploy comprehensive defenses against ballistic missile attack is the most rational way to proceed."

In any case, more than a year after Nunn's attempt to obtain from SDIO a system design to do a potentially useful job, no such proposal seems to have emerged.

It is of crucial importance that we be honest and competent in our assessment of the dangers and of means for protection against those hazards. In testimony to the House Armed Services Committee, Robert C. McFarlane described the origins of the SDI programme, detail never before revealed about the roles of then-President Reagan and McFarlane as his National Security Advisor. Reflecting on the claims made for SDI as the only way to prevent nuclear war, McFarlane testified [5]:

"...I feel guilty as a member of the community for all of my lifetime as an adult that I have not been very candid with the American people about their safety."

"...When one says that it would be irresponsible for a President to propose defending missiles rather than people, I have to say that that may or may not be true by referendum if you explain it in honest terms to Americans, that there are many ways of preventing a nuclear war ..."

"...I am not sure how it would come out, but I do know that we have never had the integrity to explain it in those terms."

2.7 The Future

Here I offer a restatement of the programme previously presented to Pugwash, namely that it is feasible and desirable to:

- Reduce US and Soviet strategic offensive forces to 2000 warheads within five years of de-MIRVing, with verification aided by tags and seals.
- Seriously negotiate to cap or reduce British, Chinese and French forces to 200-300 warheads each.

- If analysis shows a serious hazard of accidental or un- authorized launch, deploy on each side a command-destruct system so that a nation may destroy its own operational missiles in boost-phase by a coded radio signal.
- Provide cooperative silo-watchers (in fact, black boxes) that will signal continuously and without possibility of falsification that the silo door has not been opened.
- Strictly comply with the ABM Treaty.
- Ban space weapons and ASAT tests.
- Bring about a comprehensive ban on nuclear explosions by a progressive reduction in number and yield of tests permitted under the Threshold Test Ban Treaty.
- Eliminate all non-strategic nuclear weapons in the context of deep cuts in conventional forces.

References

1 *International Institute for Strategic Studies, "The Military Balance 1988–1989"* (London)

2 H. M. Agnew, "What is the right SDI approach?", in *The San Diego Union*, December 1988

3 "Discriminate Deterrence". Report of the *Commission on Integrated Long-Term Strategy*. Co-Chairmen F. C. Ikle and A. Wohlstetter, January 1988. (Washington, DC: US Government Printing Office)

4 Senator Sam Nunn, "Arms Control in the Last Year of the Reagan Administration", a speech to the Arms Control Association, 19 January 1988

5 R. C. McFarlane, oral testimony to the Armed Services Committee, House of Representatives, US Congress, 17 May 1988 (excerpt of transcript available from R.L. Garwin)

Reductions in Strategic Nuclear Warheads

Frank Blackaby

1. Introduction

The subject of this paper is the reduction in the number of strategic nuclear warheads possessed by the United States and the Soviet Union. The questions can be framed as follows: in what way can academics who consider the present situation unsatisfactory help to bring about a change? Are there ways in which they can accelerate the negotiating process? Are there ways in which they can bring nearer a further stage of reductions? What are the obstacles in the way of agreed reductions? Should the main objective be the reduction in the total number of warheads, or should there be some other objective?

The central fact about the present situation is that it is absurd – though it has persisted for so long that the absurdity is forgotten. It comes naturally to physical scientists to take a global, or planetary approach to problems. Their subjectmatter is not national: national boundaries are irrelevant to the study of physical laws. The planetary picture of the present situation is this. Two governments of two sizeable areas of the world have pointed at each other – ready for launch in a very short time – unimaginable destructive capabilities. The "observed" measure of destruction is the Hiroshima measure – 15 kilotons, which produced over 100 000 immediate deaths, and about the same number of latent deaths and physically-scarred survivors. The strategic armouries of the superpowers now contain some 10 000 megatons of explosive force, in some 25 000 warheads. (This does not include the US intermediate systems which could strike the Soviet Union). The possession of this quantum of lethal power is, of course, absurd. This central fact is banal; it does not make it any the less central. It has to be the starting-point for any thinking about this problem, and indeed it should determine the nature of a great deal of the subsequent analysis. There are a number of reasons (discussed below) why some analysts tend to forget this central proposition, and preoccupy themselves with the second-order issues.

There are no experts in nuclear war. Expertise can only be derived from the study of past experience; there is no past experience of nuclear exchanges. Those who have expertise in the study of conventional war are probably not the best people to think about nuclear war. Bernard Brodie's classic dictum of 1946 – that the role of the military is no longer to win wars, but to prevent them – has still to be absorbed into a good deal of military thinking about nuclear exchanges.

2. Analytic Approaches

"Human beings cannot bear very much reality". (T. S. Eliot). This is certainly true when the reality is an enormous lethal absurdity. It is uncomfortable, and in some ways intellectually unmanageable, to think that one lives under this vast overhang of instant lethal power. One reaction is psychological withdrawal – and strategic analysts are not immune. There are various ways of taming, and drawing the sting from, large and unpleasant facts. The main technique is to forget the lethal properties of the objects being considered – euphemisms, like the term Peacemaker for the MX missile, help. Then they can be manipulated without mental distress.

A comparison with the Theatre of the Absurd helps to explain the mental processes of withdrawal. Examples of the Theatre of the Absurd start with a totally ridiculous situation – a rhinoceros in a tube-train, or whatever. Then within the greater absurdity an enclave of rationality and normality is established. The absurdity is cocooned in normal everyday dialogue and ordinary routine exchanges. Hence the comic effect. The Theatre of the Absurd serves as a model of the way in which human beings deal with something unimaginably absurd: they try to establish an area of internal rationality and normality in which the greater absurdity is ignored and eventually forgotten.

One of the ways in which the strategic confrontation is cocooned, made mentally manageable, and generally domesticated, is by treating the whole subject as an exercise in games theory. Many who do not explicitly use games theory nonetheless adopt some of the same basic assumptions – that this is a two-player competitive and adversarial process, a zero-sum game where the single object of each player is in some sense to win, and defeat the other side. Each player is wholly concerned with the game; has no other concerns; and is ready at any moment to exploit any change in the structure of pieces which could be turned to his advantage. The strategic weapons are the pieces on the board. This type of model, with variations, is implicit in a good deal of strategic analysis.

This way of thinking about the strategic confrontation is defective in a great many ways:

- The objective is wrongly defined. There is no concept of security (except as total defeat for the other side). It is a zero-sum game – gains for one side are losses for the other. There is nothing in the model to show the immense gains to be derived from stopping the competition – from abandoning the game altogether. (The concept of security is discussed below).
- The players have nothing else to do – the game represents the total militarization of international political relations. Each side is totally and wholly concerned to exploit any advantage arising from the changing relative structure of the pieces.
- To the games player, there is no presupposition that reductions on both sides improve the situation in any way. Indeed the position with reduced pieces

on either side may well be seen as more risky and dangerous than the initial situation with a full complement of pieces.

- The strategic weapons are simply pieces on the board, their lethal consequences forgotten. No question is asked about how the pieces got there, or why there are so many.

Of the many defects of the adversarial model, perhaps the most important is its impoverished idea of the nature of security between nations. It implies that security is solely a function of the relative military structure of potential enemies. Security is something which can be bought by military expenditure, and more military expenditure buys more security. Yet the nature of security – and indeed the emergence of security – can be studied in examples drawn from the real world. Whereas with nuclear war we have no examples which can be used for study, for international security such examples now exist. There are large regions of the world where – as between the states in those regions – there is international security, in the sense that the whole idea of the use of force for settling disputes within the region is off the map of possibility. This is true now within the European Community; it is true in Scandinavia; it is true as between Canada and the United States; and so on. Security within those regions is not the consequence of a military balance. There is no military balance between Norway and Sweden; the military deployments within the countries have become totally irrelevant to the security structure within the region. Nobody looks at the force-to-space ratio along the Franco-German border. Security is reached when political relations are demilitarized. Increased world security requires an extension of this process.

The zero-sum game is a bad model for thinking about security. A much better model for thinking about desirable military policies is the standard *policy* model. *Policy* consists of action directed towards certain ends: that applies to military and arms control policy as much as it does to health or educational policy. There are two requirements for good policy. The first is a clear and precise definition of the desired end, or objective, of policy. There may well be intermediate objectives which serve the final objective. The second is a full understanding of the system within which policy operates – which is required for knowledge of the consequences of certain actions. Here, the objective is *security*, in the sense of the absence of any risk of military attack. The knowledge of the working of the system, and the way in which such security might be obtained, is greatly helped by the fact that *oases* of such security now exist, and it is consequently possible to study the way in which security in those regions was obtained, and what conditions are necessary. Perhaps the most potent phrase to explain the emergence of these oases of security is the phrase *the demilitarization of political relations*. In the political relations between the countries of the European Community, military considerations are irrelevant. When there is some dispute between Great Britain and France about sheep imports, nobody calls the Chiefs of Staff into the Cabinet discussion.

Different ways of thinking about security lead to different conclusions about arms control policies. If the adversarial approach is dominant, then arms control negotiations can become attempts to obtain military superiority by other means. It is a very different matter if the final objective is seen to be the kind of security which exists now between France and Germany.

3. Implications for the Negotiations about Strategic Weapons

What should be the objective of the negotiations? There have been some suggestions that reductions do not matter – that there is not much to choose between a confrontation of 10 000 warheads on either side and a confrontation of 5000. This is the *games theory* view: there has been an equal exchange of pieces, but otherwise the nature of the competition is unchanged. This ignores all the political consequences of a successful agreement on reduction: the progress which such an agreement makes towards reducing the importance of military issues in relations between states. Such an agreement would be one step in the demotion of the importance of nuclear weapons.

It would begin the displacement of nuclear weapons as the central symbol of state power. It would (all being well) put an end to the absurd search for essentially meaningless nuclear weapon superiority. One of the main justifications for the multiplication of nuclear weapon systems has been the belief that it was all-important not to be seen to be inferior – with the unstated implication that it was also very important to be seen to be superior. This view was expressed by Eugene Rostow, the one-time head of the US Arms Control and Disarmament Agency:

> "*The present state of the nuclear balance is a pervasive and insidious political force deeply affecting political attitudes throughout the West.*" [1]

and again on another occasion:

> "*The nuclear weapon is primarily a political, not a military force – a potent political force, generating currents of opinion which are transforming our world ... I believe the risk of nuclear war is far less today than the risk that the unity of the West will be destroyed and the West reduced to neutrality by psychological and political pressure emanating from the nuclear balance.*" [2]

The process of reduction should begin to put an end to this belief.

A US-USSR agreement on reductions would at least remove one argument used by the near-nuclear-weapon states: ... that the nuclear powers express immense concern about horizontal proliferation, but have done nothing significant about vertical proliferation. How effective the removal of this argument might be is somewhat doubtful, because it has come so late. It would, however, serve to strengthen the opposition in near-nuclear weapons countries, in so far as such an opposition exists. It is a downward and not an upward pointer for nuclear weapon systems.

Some of those who have become particularly concerned about the risk of accidental nuclear war tend to underplay the gains which could be made simply by reducing numbers. First, it is clear that at every point where it is physically impossible for a separate decision to be made on firing a nuclear weapon, at that point there is a separate risk, namely of an individual or group of individuals taking what might be called a pathological decision to fire a warhead. Sea-launched cruise missiles are a good example. If a large enough group of those on board a vessel carrying these weapons decided to fire it, central authority would have no way of stopping it. Obviously, if the number of vessels carrying these weapons is reduced, then the risk of a *pathological* accident is also reduced. Where there is a central system of some kind which relays instructions to possible firing points, the risk of an accident is not confined to the central system; there is some risk in each of the separate links to each firing point, and an increase in the number of firing points increases the risk. The larger the network, the greater the risk – it increases, though it may not necessarily increase proportionately.

In analogous civil systems, there has never been any question but that the risk of an accident increases with the size of the system. With telephone systems, the risk of a crossed line – not dissimilar to the possibility of a wrong message being received – increases with the number of lines installed. With an electricity distribution system, again the risk of accident increases with the size of that system.

One of the important elements in accident risk is the fact that nuclear weapon systems are kept on a high state of alert. Here again, it is appropriate to consider the likely concomitant consequences of an agreement on reduced numbers. It would be seen as part of a process in which the whole idea of deliberately-initiated nuclear war becomes less likely, and the alert status of various weapon systems would probably be reduced.

4. Risks of a First Strike

The obstacles to agreement do not lie in the technical military field at all. There is a huge range of possible agrements, all of which would increase the security of both sides by reducing the military component of political relations. There is not some narrow path of reduction which is the only one which avoids the risk of a first strike by one side or the other. No United States or Soviet President (assuming that they stay sane) would order a first nuclear strategic strike against the other side, either now or during any of the very wide variety of routes to reduce the present deployment to a half.

Again, the picture of the threat of a first strike derives from the games theory model – that each side has as its main and only objective to conquer the other side in some sense, and each side would immediately take the risk if there seemed to be even a slight chance of successfully eliminating the other side with nuclear weapons. The whole idea of sane politicians risking everything on a slight chance of a successful first strike, knowing that the consequences of anything going wrong would be the destruction of their nation, is absurd. Nuclear weapon policy must have as its assumption that the rulers on either side are adequately sane. (There is no structure of deterrence which would be effective against the insane).

This does not prevent some people from putting forward arguments about the risks of a first strike on the way down. Here is one such presentation:

> *"A back-of-the-envelope calculation by an official starts with the START goal of 4900 warheads allocated to each side. Of those, 1100 would be allocated to ALCMs on bombers; if each bomber is allocated 10 ALCMs, that means only 110 bombers. Since only one-third of bombers are on alert, the United States would have only 37 bombers on alert. The others take 12 hours to be mission-ready. As for the ICBMs, 50 MX will take 500 warheads, and 100 Minutemen III will take 300 warheads – each in a fixed silo, on which the Soviet Union could theoretically target three warheads. That leaves 4100 warheads for Submarine-launched Ballistic Missiles. Each Trident boat caries 24 missiles with 10 warheads, or 240 total warheads, so only 15 boats would carry the remaining warheads. Only 60 per cent of these would be on station, hunted by Soviet SLBMs, twenty times their number."* [3]

The unwritten assumption behind the implied scenario is of an adversary combining wickedness and lunacy, whose sole concern is to look for the faint chance of a successful first strike, and who would take it on the extremely remote chance that it might come off without total devastation of the Soviet Union. It is this kind of thinking which has helped to create the absurdity of the present situation.

Thus, the *window of vulnerability* was one of the arguments for the present US programme of *modernizing* US strategic weapon systems during the past decade. The argument was that the Soviet Union could launch a first strike against US land-based intercontinental ballistic missiles, and that the US would be inhibited from retaliating against Soviet cities with any of its other nuclear weapons systems, because that would bring further destruction on the United States. We were invited to believe that a sane Soviet leader might launch that first strike against a small proportion of US strategic nuclear capability, on the off-chance that the United States might be inhibited from retaliating. This was absurd. Once again, any system of deterrence has to assume some degree of rationality on the other side, because the totally irrational would not be deterred by anything.

There is extant a draft START Treaty with a large number of square-bracketed alternatives – on the permitted range of this or that missile, and so on. Either side could in fact accept in total the square-bracketed proposals of the other side, with absolutely no risk of a first strike (so long as rulers stay sane) – which is the only deliberate risk that matters. Even on such major issues as SDI and sea-launching cruise missiles, that still remains the case.

It is a temptation to go through the square-bracketed draft, and to produce a compromise. This would be a misidentification of the nature of the obstacles to an agreement. The devil is not in the detail. If the two sides were told "reach any agreement, so long as there is no reasonable risk of a first strike" they could complete the negotiations in a week. The problem does not lie there, but in the structures of thought of the main players, perhaps particularly on the US side. (There seems at present more acceptance of new ways of thinking about security on the Soviet side).

There has been extensive discussion of the relative destabilizing character of different nuclear weapon systems – that in the end, so long as it is considered that a nuclear deterrent structure has to be maintained, a small number of single-warhead invulnerable missiles should be preferred; and that in the process of reduction it would be better to get rid of vulnerable multi-warhead missiles; and propositions of that nature. However, these are second-order issues; the main requirement is to get a reduction, and all the political benefits which would accrue from that. The risk of a first strike from some less favourable pattern of reduction is not real.

5. Obstacles to an Agreement: Old Ways of Thinking

The adversarial model, the zero-sum model, and the idea that there is political mileage to be gained from being seen to be superior – these are still potent in a good deal of thinking about security. They give rise to a negotiating stance where the objective is to win as many points as possible – each win being a loss to the other side. It is essentially a military competition continued by other means. The objective is to come away with a clutch of concessions, which are then paraded like captive prisoners.

The idea that it is not a zero-sum game, and that there is a considerable common gain to be made from an agreement, is still not a powerful idea: it seems to exert little influence in the West. There is very little stress on the potential economic gains from reductions in the scale of the military confrontation.

More generally on security, the idea that it is something which is bought by military expenditure is still strong, as is the idea that security consists of being militarily more powerful than potential adversaries. There is still only weak acceptance of alternative ideas, such as that in the long run security can only be obtained in common with – that is, in agreement with – a potential adversary. Further, in the even longer run, there is little understanding of the nature of security which now exists, in some cases between countries which were enemies less than half a century ago, but where military capabilities have become totally irrelevant in the relationship. It is proving very difficult for the West to adjust to the possibility that the Soviet Union might not for ever be the enemy. It is essentially old ways of thinking about security which are the obstacles now to progress in reductions in strategic weapons.

Another obstacle to a successful outcome to negotiations is the absence of public concern and public pressure. Politicians like to give their time and attention to matters which, if properly resolved, will win them votes. If there is little public interest in any set of arms control negotiations, then the issues raised by those negotiations will be given no priority. They can drag on for years – as indeed most arms control negotiations have done – and their failure loses no votes at all. The Mutual Balanced Force Reduction (MBFR) negotiations are of course the prime example – continuing for fourteen years with no result. It is doubtful whether any politician lost a single vote as a consequence of that failure.

Once public opinion has been roused on an arms control issue, then it becomes a very different ball-game. The best current example is that of West Germany. The commonly-accepted view for the last few years has been that the peace movement in Germany was dead: there were no more mass demonstrations, once the INF Treaty had been agreed. The observers who said this had failed to notice how far public opinion had been influenced against nuclear weapons on German soil, and how far the idea of the Soviet Union as the inevitable enemy had been eroded. West German politicians found that it would lose them votes if they supported the introduction of new short-range nuclear weapon systems into Germany, and that it would gain them votes if they opposed that idea. West Germany's whole stance on arms control issues was changed by the pressure of public opinion.

One of the standard criticisms of those who try to stimulate public interest in arms control issues is that this serves as a pressure on Western Governments, but not on Eastern. This is no longer true. In Hungary, Poland and the Soviet Union groups can be formed which are genuinely independent, and which can examine critically their Government's arms control proposals.

6. Conclusion

The question at issue is that of attempting to accelerate the progress of disarmament in the strategic nuclear weapons field. This requires an assessment of the nature of the obstacles. These lie – it is argued – not in the military technicalities – there exists a very wide range of possible agreements, all of which would enhance security. The difficulties are perhaps more deep-seated – in an inadequate understanding of the nature of security. Possibly, therefore, those who want more rapid progress should concern themselves less with the negotiating detail, and more with a general critique of old ideas about security – for instance that it can be acquired unilaterally by attempting to establish military superiority.

There is need for a much wider understanding of alternative ideas about security: that, initially, progress can only be made in common with a potential adversary. This means abandoning the adversarial, zero-sum game approach to negotiations, and treating them more as a cooperative approach to obtain for both sides the huge advantage which would follow as political relations become demilitarized. In the end, security is a state where relative military deployments become irrelevant.

Rapid progress is unlikely without public pressure. The present example of West Germany shows that it is possible to raise public consciousness of these issues. President Dwight Eisenhower's dictum is here relevant: that the time might come when people demand disarmament so strongly that politicians have to get out of the way and let them have it.

References

1 *United States-Soviet Relations*. Hearing before the Committee on Foreign Relations, US Senate, 98th Congress, first session on Arms Control Resolutions, Part 2 (US Government Printing Office, Washington, DC, 1983), p. 191

2 Ibid, page 210

3 Arms Control Reporter, section on Strategic Arms Limitation Talks, 15 April 1988, p. 611. C. 53, *Institute for Defence and Disarmament Studies*, Brookline, MA, USA

The Role of "Glasnost" in the Future of the Arms Race

Milo Nordyke

1. Introduction

The "Cold War" construct that has evolved in the West to describe the relationship between the Western Powers and the Soviet Union is now over 40 years old. With the exception of a few ripples driven by Khrushchev's modest reforms and Brezhnev's "detente", it has been essentially static during that time. However, momentous things are happening in the Soviet Union. Some of them may well disappear in a few years; others will persist and change the Soviet Union forever. The construct we have come to rely on for our thinking about the Soviet Union must change to reflect the new realities in the Soviet Union. But we should be careful to ensure that the changes are gradual and are tempered by experience with the new "Gorbachev-style" Soviet Union.

The key question for the United States and the West is "What kind of a relationship with the Soviet Union do we want in 10, 20 or 30 years? What is our image of the future for this relationship?" We should be examining such questions today, so that we can encourage those developments in the Soviet Union that move in directions we want, and discourage those that do not. It is in this context that I would like to examine the unique role that one of the key elements of perestroika –"*glasnost* – the openness about public affairs in every sphere of life" – is having on the US-Soviet relationship and what it could mean for the future of strategic relations between the United States and the Soviet Union.

2. The Arms Competition

The most important cold war issue is the arms competition between the USA and the Soviet Union that has seen the number and types of nuclear and conventional weapons systems build up to truly enormous quantities, with continuous major investments in qualitative upgrading of those systems. To plan our future, it is useful to look at the history of the "arms race" over the last 40 years and try to identify the factors that have driven it. Of course, one can only speculate about the relative importance of the various factors, particularly on the Soviet Union's side.

From the US perspective, there have been three main engines driving our national security planning and weapons procurement policies: Soviet declaratory policy regarding their political and military philosophy and intentions; the concrete actions by the Soviet Union in terms of hardware testing, procurement and deployment as observed by the USA; and *worst-case* planning regarding present and future Soviet capabilities, generally based on mirror-imaging US technology.

While the first two factors have provided the motivation and political force for implementing US policy planning, they have not had a major impact on the decisions regarding developments of specific weapons systems. Declarative policy statements by the Soviets have been useful for developing the political base required to develop the West's security system, but they are generally too diffuse for deciding what to do about individual elements of our military arsenal. Our observations of Soviet hardware development and deployments generally have come too late in the weapons cycle to be useful for US decision-making. In a real sense, it has been the third factor, worst-case planning, that has driven US decisions on development and procurement of strategic systems.

The rationale for worst-case planning is generally based on the fact that we do not know what the Soviet plans and capabilities actually are. Therefore, to avoid being surprised technologically or militarily, we have assumed that the Soviets can and will do anything we can do and, therefore, we should do it first. While this so-called "technological imperative" policy has received a good deal of criticism in some circles over the last 10–20 years, it is a quite rational response to the situation in which the USA has found itself. The Soviet Union has been very successful in denying the West information on their research and development activities, and their plans for future weapons systems. This lack of information has provided fuel to this engine and has made it very difficult for people uncomfortable with the "worst-case" planning approach to argue for more solid and realistic national security planning. It has led to a number of embarrassing and costly overestimates of Soviet capabilities, such as the "bomber gap" and the "missile gap' in the fifties and sixties. It has also led to the USA being the first to introduce virtually every new technological innovation into the arms race, from MIRVing to stealth aircraft.

From the perspective of the Soviet Union, I believe the picture is much different. The historical Soviet obsession with security from foreign attack is well documented. Over the past 40 years, this has led to an apparent political willingness to support the development and deployment of a wide variety of strategic systems, with few questions as to their utility or specific role in the defence of the Soviet motherland. It is clear that under Gorbachev this is changing, and that much harder questions are being asked of the Soviet military than in the past.

Decisions by the Soviet military on future directions for research, development and production of new weapons systems are undoubtedly based in part on what they see happening in the USA. From the viewpoint of the Soviet military, the USA is essentially a transparent but unpredictable society, at least from the viewpoint of defence planning and strategic policy. At the early stages of development of new technologies and weapons systems in the USA, there is an active public competition among sponsors for acceptance and support. From the Soviet point of view the problem is one of deciding which of these ideas will be accepted as elements of the US strategic planning and development, as well as selecting which are worth pursuing within their own R & D community.

Another characteristic of the US defence policy is its intermittent nature. Because the responsibility for formulating and implementing US defence policy is split between Congress and the Executive branch, and because the time required for the development and procurement of almost all modern weapons systems is much longer than the "political rhythm" of both these branches, almost all such systems experience significant slow-downs and delays, and in many cases cancellations, between the decision to begin development and the actual deployment of the systems. Numerous examples in recent history come to mind such as the B-1 bomber, MX, or the small mobile missile (Midgetman).

Soviet defence planners watch this process very carefully, develop their plans taking into account US decisions, and proceed with their research and development. However, within the Soviet system, once a decision is made to proceed, it is much more likely to proceed unhindered than in the USA. Therefore, even if their development and procurement system were less efficient than our own, it should not be surprising that they sometimes deploy new weapon systems well before the comparable US systems appear. The most recent example of this is the deployment of mobile launching systems for both intermediate range (SS-20) and strategic missile systems (SS-24s and 25s).

3. The Role of Glasnost

In the past year or so the new *glasnost* policy has led to significant changes in many different aspects of our relationship with the Soviet Union. I would like to focus on just three that, taken together and extended into the future, could well have a significant effect on the arms race. As a common thread, they all involve major changes in the amount and kinds of information available about the Soviet Union and how it operates glasnost.

Treaty Verification Activities. An area of enormous significance in which *glasnost* has had a major impact is the opening up of major elements of the Soviet military establishment through arms control agreements. The INF Treaty, signed in December 1987, with its broad-ranging on-site verification provisions was a milestone in this regard. The Memorandum of Understanding, exchanged even *before* the Treaty was signed and ratified by either country, contained details of the Soviet military infrastructure and weapons deployments that were simply mind-boggling to anyone familiar with such matters over the last 40 years. The on-site inspection procedures will result in repeated visits to over a hundred Soviet military bases throughout the Soviet Union and Eastern Europe over the next 13 years. It also established a US portal monitoring facility outside a Soviet missile assembly plant at Votkinsk, to monitor and inspect SS-25 mobile missiles as they come out of the facility, to ensure that the Soviet Union is not secretly producing SS-20s banned by the INF Treaty.

But even more important than the information that will flow from these activities, in the course of these inspections many hundreds of US and Soviet military and civilian personnel will be interacting on a one-to-one basis, learning how each other think and act. For the first time, we are beginning to obtain hard information about the Soviet military system: how well things really work or don't work, what the quality of the military infrastructure really is, what the military chain of command is, and how it responds to problems.

In another area, the USA and the Soviet Union conducted last summer the Joint Verification Experiment (JVE) which was part of the negotiation process for enhanced verification of the Threshold Test Ban Treaty (TTBT). The JVE consisted of an on-site yield measurement made on a nuclear explosion at each country's nuclear weapons test site by scientists and technicians from the other country. Although they were only meant to be demonstrations of the technology required for such measurements if they were carried out on a more extensive basis, they represented unprecedented access by each country to the other's nuclear test site. Over 150 Americans travelled to the Soviet nuclear test site near Semipalatinsk, a top-secret area of the Soviet Union where they participated in the preparations and conduct of a nuclear test. A somewhat smaller number of Soviet scientists and technicians came to the Nevada Test Site. A great deal of useful information was gained on the characteristics of the test sites by both sides, how operations are carried out, who the people are that are involved, and how technical problems are resolved.

As part of this operation, the USA drilled, logged and instrumented an 1800-foot-deep hole at the Soviet test site. In order to carry out this operation the Soviet Union permitted the USA to fly some seven C-54 military aircraft from the USA to the airfield at Semipalatinsk and back, to deliver the drilling and logging equipment to the test site.

While the character of the final agreement to enhance the verification provisions of the TTBT is not yet resolved, what is clear is that any future agreements in this area will provide for many more such operations, if not a virtual presence of a large contingent from the other side at each other's nuclear test sites. All of these events represent breathtaking departures from Soviet behaviour of the past 40 or more years, and have provided significant new insights into the operational practices and capabilities of the Soviet military establishment.

If and when a START Treaty is negotiated, it is clear that it will have verification provisions that go far beyond those in the INF Treaty, by including surveillance of production facilities, visits to ICBM deployment areas, aircraft bases and, very probably, naval ports that are presently strictly off-limits to any foreigners. The information and contacts experienced with the INF and TTBT verification activities will be multiplied many times over after several years of living with a START Treaty.

The Conventional Forces in Europe (CFE) negotiations on military forces in Europe, which started in March 1989, will provide still more opportunities for interaction between the Soviet and US military. If the standard of verification provided in the INF Treaty and expected in the START and TTBT is met, the level of military interaction and information exchange will be a major escalation over that which has already occurred, and will extend it to allied forces as well. President Gorbachev has already proposed that an exchange of military data be carried out *now* with verification of those levels by on-site inspection to be carried out *during* the negotiations.

I am sure that it has been very difficult to get the Soviet military to accept this opening up of the military infrastructure. Of course, essentially all of this type of information about US military forces has been available to the American people (and presumably to the Soviet Union) for many years. We believe that the societal costs of keeping such information secret are not commensurate with the benefits from a national security viewpoint. In fact, most people believe that there are disbenefits from being unnecessarily secretive. I would also like to argue that the excessive secrecy practiced by the Soviet Union over the past 40 years, and in particular by its military, has been counter-productive to Soviet ends, since it not only provided a shield for the incompetent, but, more important from the arms race point of view, it provided a rationale for the US worst-case planning. Gorbachev appears to recognize that fact, also, and has so far been quite successful in introducing *glasnost* into the military world. Of course, this is a reversible process and we must be careful to set a high standard for observance of such agreements.

Exchange of Military Planners. A second area with the potential for a significant impact on the defence planning and threat assessment of both countries, is the exchange of visits and consultations between senior representatives of the defence community. During the past year, the US Secretary of Defense and Chairman of the Joint Chiefs of Staff carried out an exchange of visits with their Soviet counterparts – the Soviet Defence Minister and the Chief of the Soviet General Staff. If such visits are continued and hopefully expanded to include one or two lower levels of defence planners, they will promote a much deeper understanding by both sides of how the other thinks and a greater reality in assessing the threat we both face.

Another activity in this general area that could become much more useful in the future is the interaction of defence and arms control specialists. Before Gorbachev, such interchanges were quite barren exercises, largely because the Soviet participants almost always took positions espousing and defending the official government position. Such interchanges are becoming more interesting as a small measure of diversity is starting to creep into the arms control community in the Soviet Union, although these discussions are still limited by the small number of Soviet participants who feel comfortable to express personal views on subjects related to state security. In addition, it is clear that all serious discussions of such subjects before Gorbachev were limited to military or very high-level political circles. Thus, no significant group of non-military specialists existed in the Soviet Union comparable to the so-called "arms-control community" in the West. Developing such a community in the Soviet Union will take a number of years of *glasnost* and, to a large extent, will be dependent on the intellectual demands of their Western counterparts for candour, resourcefulness and honesty.

Application of Glasnost to General Information Policy. A third area of change really is the general application of *glasnost* to the information policy of the Soviet Union. *Glasnost* represents a fundamental change in how the Soviet Union handles information. Only time will provide us with an understanding of how far-reaching and how lasting this change will be. Although its primary justification was as a contribution to the rebuilding of the Soviet economy and the "democratization" of Soviet socialism, one of its most important impacts may well be in the area of foreign affairs and strategic relations.

Recent events provide two examples. Over the past three or four years, the Soviet Union has been actively working with several American groups on the deployment of seismic stations throughout their country. Prior to this initiative, there were no seismic stations in the Soviet Union – the largest land mass on the face of the earth – that routinely published a record of seismic events. For some reason, seismic data were regarded as not being appropriate for public release. This represented a very large hole in the world-wide network of seismic stations and severely limited the world's knowledge about seismicity in that part of the world, particularly for the smaller seismic events that are of particular concern from an arms control point of view.

An agreement has recently been signed between a private US organization, the International Research Institutions for Seismology (IRIS), and the Soviet Academy of Sciences that will permit the establishment of between 10 and 15 high-quality seismic stations in the Soviet Union. The data from these stations will be collected and published by IRIS. The primary motivation for this agreement is to complete a world-wide seismic network being deployed by IRIS, but these data will also provide a solid base for analysing a number of the key technical problems involved in considering further limitations on nuclear weapon testing.

A second example is the recent earthquake in Armenia which received quick and widespread publicity throughout the world. This was in stark contrast to the total secrecy that surrounded a similar earthquake with similar tragic consequences some years ago. The publicity for the recent earthquake led to world attention and an unprecedented outpouring of assistance to the Soviet Union in dealing with the human dimensions of that tragedy. Not only did the world learn about the earthquake's destruction and human misery, it also learned that the Soviet Union had no significant civil defence capability to deal with such area-wide disasters. Such a capability has long been alleged for many years by some of our defence planners as the basis for our worst-case estimates of Soviet recovery from a nuclear attack. The Russian people and the Soviet leaders also learned that the world outside the Soviet Union cared about them as human beings – that being vulnerable made their country more a part of the world community than being strong. A number of Soviet spokesmen, including Foreign Minister Edouard Shevarnadze made a point of commenting on this aspect.

4. Conclusions

All of these developments – the opening up of the Soviet military infrastructure and the interaction of military personnel, the high level exchanges between US and Soviet defence planners, and the application of glasnost to the general information policy in the Soviet Union – if continued and expanded in the future – will have a significant long-term impact on the military planning and decisions regarding weapons systems development, procurement and deployment for both countries. But the greatest impact will probably be for the United States. Instead of making worst-case assumptions about the capabilities of each other's forces, strategic military decisions can be made on more realistic assessments of their strengths and weaknesses and actual capabilities and plans. To the extent that Soviet openness has this result, both countries will benefit from more reality-based planning with reduced military requirements. Hopefully, it will have served to reduce and fundamentally alter this common enterprise we call the arms race.

The critical question we should be addressing is whether there are other activities the USA can initiate which would encourage this process. Are there interactions we can propose which will challenge the Soviet Union to open the doors even more than they have so far? A useful step in this direction is a recent publication by the Johns Hopkins Foreign Policy Institute *The Future of US-Soviet Relations: Twenty American Initiatives*. The book describes some 20 ideas for developing cooperative ventures which, as a common theme, would provide unique benefits for both countries. Suggested initiatives range from an American college in Moscow to converting nuclear missiles for peaceful uses.

One suggestion, by Jack Perry, for an advisory council on American-Soviet relations could be particularly relevant to strategic planning. His concept calls for a select and prestigious body appointed by President Bush that would assess the state of US-Soviet relations and provide advice on foreign policy issues based on such assessments. There are obviously many concerns that should have to be addressed before implementing such a suggestion, but it is a useful idea for dealing with the persistent problem of US inconsistency in foreign policy.

An extension of this idea would be for the establishment of a high-level cooperative strategic planning group made up of senior military and civilian members of the US and Soviet defence communities. There are great differences between our two countries. Our disparate geography, resources, and history impose very different requirements on our respective defence establishments. Understanding such differences and their impact on defence decisions requires extended discussions by openminded and knowledgeable experts. The suggestion here is for a forum dedicated to discussion and not negotiation on these issues. Out of such discussions could come a better understanding of the basic motivations and world-view of the other side. Such understandings are fundamental for making more rational long-term decisions regarding the defence policies for both countries.

It would appear very probable that Gorbachev will be in charge for some period of time, be it two years or ten years. We don't know and he doesn't know. Nor do we know who or what type of regime will follow him. During his tenure, we have a "window of opportunity", during which we can try to encourage and expand the openness and flexibility he has demonstrated so far with the measures described above. To the extent that we respond to that openness with positive measures which demonstrate the benefit to the Soviet Union of such a policy, we will be helping to ensure that the regime that succeeds Gorbachev will follow his lead.

Accidental Nuclear War

Kurt Gottfried and Alexander George

The 18th Pugwash Workshop on Nuclear Forces was held in Pugwash, Nova Scotia on 18–20 July 1989. Its purpose was to understand better the factors that could contribute to accidental nuclear war, and in particular to suggest what could realistically be done to ameliorate the risks in the near or foreseeable future. Participants comprised 30 specialists from 10 countries, drawn from various disciplines and professional backgrounds in order to provide a multi-dimensional and reasonably comprehensive approach to this problem. The following account draws upon and attempts to faithfully reflect material in working papers prepared by participants and the discussions. The accuracy of this report is the responsibility of the rapporteurs alone; it has not been reviewed by other participants and would not necessarily be fully endorsed by them. Given the range of issues covered in the Workshop and the richness of the papers and discussion, the rapporteurs have found it necessary to prepare a longer, more detailed report than is customary for Pugwash workshops.

1. Accidental and Inadvertent War: Definitions and Scenarios

Members of the Workshop were not able to agree on precise definitions of "accidental war" and "inadvertent war". (What "accident" and "inadvertent" war have in common is that neither is a *premeditated* war – i.e., the classical "preventive war". In the modern era, the preventive war option is associated with the hypothetical possibility that one of the superpowers might plan and initiate an unprovoked nuclear war, most likely under peacetime conditions in order to achieve surprise, in order to destroy its opponent's war-making potential and to eliminate or sharply reduce its ability to compete in the international arena. Preventive war was excluded from our discussion.) More importantly, however, there was substantial agreement on *scenarios* that could possibly result in an unintended and unwanted nuclear war and on a variety of factors that could possibly contribute to such wars. Preventive measures were identified that require and deserve analytical assessment and consideration by policymakers.

"Accidental war" received much more attention in the meeting than "inadvertent war". "Accidental war" can be defined as one that occurs as a result of a wholly erroneous or significant misinterpretation of tactical warning that the opponent has launched a substantial nuclear strike which leads, in turn, to "prompt launch" of major nuclear retaliation. (We use the term "prompt launch" in this report to include both "launch-on-warning" of an attack that is underway and "launch-under-attack".) Included in this definition is the possibility that what is in fact a minor nuclear attack is misjudged in early attack assessment to be a substantial one.

"Inadvertent war" differs in that it is not tied, as is "accidental war", to misinterpretation of tactical warning or to erroneously inflated attack assessment. Rather, according to one scenario of inadvertent war, nuclear war occurs because one side deliberately decides to initiate *pre-emptive* nuclear war on the basis of its assessment of *strategic* warning that the opponent has decided and is making serious preparations to launch a major nuclear strike in the very near future but has not yet set his attack in motion. Strictly speaking, by definition, "pre-emption" is undertaken on the basis solely of strategic warning *without waiting for tactical warning that the opponent's attack is already underway*. (In practice, of course, a pre-emptive war can be initiated on the basis of a combination of strategic warning and some partial but inconclusive tactical warning indicators, or on the basis of a mixture of equivocal strategic and tactical warning).

Much of the discussion focused on factors (technical and human) that can contribute to different accidental and inadvertent war scenarios. Such factors can combine in different ways to cause war, i.e., there are different paths that can lead to accidental or inadvertent war. For each of the technical factors we add a brief discussion of some of the preventive or prophylactic measures that were suggested by participants. Following that, we discuss some of the human factors that could contribute to either type of war: the impact of crisis-induced stress, organic illness, cognitive dynamics, disturbed emotional states, and sleep deprivation. Finally, since crisis is a major context in which accidental or inadvertent war might occur, we offer some observations about the importance and modalities of crisis avoidance and crisis management, and call attention also to the importance of the state of the overall US-Soviet political relationship as a factor affecting the likelihood of unwanted nuclear war.

2. Accidental War

Crudely put, accidental war can occur as a result of an action (or actions) that triggers nuclear war but was not properly authorized by the National Command Authority (NCA) or by a legitimately pre-delegated command. One such "trigger" will be considered here: accidental (and/or unauthorized) launch of strategic nuclear weapons. In addition, accidental war may occur as a result of a systematic failure of the warning and command system.

2.1 Accidental War Via Accidents at Low Alert Status

Should an accidental launch (unless otherwise stated, this includes subsequent detonation on a target of value) occur when both sides are in a peacetime posture, or at a low alert status (as in the Middle East Crisis of 1973, when the USSR did not raise its readiness and the USA was at DEFCON 3), most participants believed that the likelihood of serious consequences is minimal. It is not obvious that this confidence is justified, however, for the following reasons.

First, it appears unlikely that a breach of the complex safeguards (at least in the US strategic command system) could occur only at the level of a single silo or even one Launch Control Centre. Therefore, an appreciable number of ICBMs might be fired in the event, to which low probability is attached, of an accidental launch. Second, by the same token, accidental launch from a ballistic missile submarine might involve more than one missile, and perhaps even the full complement. Third, as most strategic missiles are MIRVed, such an accident could involve casualties that dwarf human experience. Hence, even if an accidental launch did not trigger a "launch-on-warning" or "launch-under-attack" response (the two are referred to in this report as "prompt launch" since the distinction between them is not clear-cut) – the absence of such a response to accidental launch may be quite likely in peacetime since it might then have a high threshold – it is unjustified to ignore the possibility that such an accidental launch would lead to demands for revenge, exceedingly tense international relations, the fall of governments, and so on, which could lead to a war that may begin at the conventional level and then escalate.

Innovations that could reduce these risks can be unilateral or cooperative.

2.2 Unilateral Measures

2.2.1 Command-Destruct Capability (CDC) on Strategic Missiles. CDC permits the missile's owner to order destruction by radio command after launch (as in test firings). The US military is strongly opposed to CDC because it fears espionage that could compromise the ability to retaliate. On the other hand, if CDC can be turned off before launch (possibly as part of the launch sequence), so that enemy radio signals could not destroy a missile that had been launched on proper authorization, these objections seem to fall away. (As to the technical dimension, it might not be feasible to instal CDC on re-entry vehicles because of the weight penalty, but that problem would not arise for the MIRV bus or the booster). CDC therefore appears to be a very promising technical means for reducing the accidental launch danger.

2.2.2 Permissive Action Links (PALS) on All Nuclear Weapons. At this time all US Army and Air Force nuclear weapons have PALs, but that is not true of Navy weapons. To what extent PALs are installed on Soviet, French and Chinese weapons is not known to us. As for British weapons, the situation presumably is the same as for the USA. There was unanimous agreement that PALs should be installed on all nuclear weapons, irrespective of basing mode or nationality. In the case of the USA, it would be important to do so on attack submarines, most particularly on the SLCMs, and to a somewhat lesser degree on surface combatants, as these vessels are intended for operations in hostile environments. Ballistic missile submarines purposely operate in great isolation even in wartime, and are under a strict command regime, so the urgency for installing PALs on them is not as high.

2.3 Cooperative Measures

Five measures were proposed that received strong support:

2.3.1 PAL Information Exchange. The USA (presumably also the USSR and France) has invested appreciable resources over many years to developing PAL technologies. It would be in the US interest, and those of other states that have sophisticated PAL capabilities, to share this technology with other states.

2.3.2 $C^3$3I Information Exchange. SALT and START have focused exclusively on weapons, and with the exception of certain limits on radars in the ABM treaty, have ignored C^3I. Given the vital importance of command, and growing mutual confidence, the US and Soviet militaries should begin systematic discussions regarding their command systems. These could educate both sides as to what peacetime (and perhaps crisis) activities are particularly dangerous. Such discussions could also lead to an enlargement of the 1989 Agreement on the Prevention of Dangerous Military Activities to cover space activities, which should also receive serious consideration given that ASAT capabilities may be in the offing. Finally, in the long term, military-to-military discussions could facilitate agreements on operational constraints for strategic forces that would reduce risks in crisis.

2.3.3 Direct Links Between NORAD and Its Counterparts. At this time the only USA-USSR direct communication links are the Hot Line and the Risk Reduction Centres. While these are adequate for communication between the National Command Authorities in peacetime under normal circumstances, it is probably unlikely that they could transmit timely warning of an accidental launch before impact. Links between NORAD (and possibly SAC) to their Soviet counterparts should be established with appropriate procedural filters to permit such communications only under well-defined contingencies.

2.3.4 Multinational Crisis Risk Reduction Centre(s). The capabilities embodied in the US/Soviet Hot Line and the new Risk Reduction Centre should be enlarged or replicated to form a global network incorporating all declared nuclear states, and possibly a larger set that would include the threshold states.

This should be done under UN auspices, or on a multilateral basis, whichever is more expeditious and efficacious. At a minimum, the existing system should be enlarged and include China, France, and the UK.

2.3.5 Reduced Alert Status of Strategic Missiles in Peacetime. The proposal was made to place the superpowers' missile forces at a lower alert status in peacetime so as to reduce the risk of accidental launch, with the proviso that this be implemented in a verifiable manner. While no such technique was put forward short of separating warheads from boosters, it was proposed that further work on this idea should be undertaken. (One possibility is offered by mechanical gyros, which take 24 hours to spin up, but new technologies, e.g., laser ring gyros, could circumvent this measure, unless installation of new devices could be prevented with adequate confidence). Some participants expressed a basic misgiving about this concept, however, for if the alert status were then raised in a crisis, this would convey a signal of extreme threat.

2.4 Accidental War via Accidents at High Alert Status

Obviously the risk that an accidental launch would lead to war is far greater in this context. For that reason most, and perhaps all, of the above measures would then take on far greater salience. In particular, PALs on naval vessels, and especially attack submarines, would be especially important, since these would then be operating in close proximity to the forces of the adversary, and/or be conducting highly intrusive intelligence missions, while operating only under one-way communications (and in the case of submarines, possibly communications silence). In the underseas environment, given the difficulty of acquiring reliable tactical information, the chances of miscalculation and accident are extremely high. In this connection, particular concern was expressed that once American and Soviet submarines begin to confront each other during a crisis, a tactical miscalculation or accident low in the chain of command could lead to an underwater Vincennes-type incident that might trigger accidental nuclear war.

Proposed measures specific to the high alert status environment, and not yet mentioned, include:

2.4.1 Reduced Reliance (Elimination of) Prompt Launch. There was widespread agreement that the rather heavy reliance on prompt launch (i.e., "launch-on-warning" and "launch-under-attack"), apparently held by both superpowers, poses a most dangerous risk to humanity, and that every effort must be made to reduce and eventually to remove it. The reason for this reliance is that both command systems are highly vulnerable to the adversary's strategic forces, while the fixed land-based forces are also very vulnerable. In consequence, there would be pressure to launch while the command system is still functioning, which is opposed by the pressure to avoid nuclear war.

The pressure to launch promptly will only be relieved if the command systems, and to a lesser degree the forces, can survive attack and continue to function adequately. Several steps in this direction are already underway: the US Commandand-Control Improvement Program, (its essential elements are high-altitude communications satellites (MILSTAR), a radio network for transmitting emergency action messages throughout the USA, and ground-mobile command centres and communication nodes; all are designed to function in a nuclear environment), which is creating a command system that can survive a significantly heavier attack than its predecessor; and START, which will reduce the arsenals by about a third; and deployment of mobile ICBMs. But these will not suffice, though they will presumably raise the threshold for consideration of prompt launch. To attain the goal, deeper cuts will be required to allow ongoing improvement in C^3I technologies to win the race against the forces that could liquidate command; and to create forces that

cannot be destroyed by pre-emption. An important improvement would be more accurate and prompt attack assessment. This might be a (the only?) benefit of the SDI programme!

2.4.2 Operational Constraints on Ballistic Missile Submarines. To reduce the danger of surprise attack, and therefore the reliance on prompt launch, constraints on ballistic missile submarine operations have been proposed repeatedly, and were endorsed by all. These would forbid patrol by ballistic missile submarines within a specified distance of, say, Washington and Moscow, or perhaps in areas having a broader geographic definition. Such an agreement can be cooperatively verified. (See *Crisis Stability and Nuclear War*. Kurt Gottfried and Bruce G. Blair (New York: Oxford University Press, 1988); pp 302–3).

2.4.3 ASAT Arms Control. ASAT arms control was also strongly endorsed for similar reasons. It should be noted that unless very high performance directed energy weapons are deployed, high-altitude satellites vital to strategic command are safe from surprise attack, as an attack by rockets would give several hours of tactical warning. On the other hand, attacks on low orbit satellites, which are important in crisis and conventional war operations, could lead to escalation.

2.4.4 Remove Ground-Based Tactical Nuclear Weapons. Ground-based nuclear weapons, especially nuclear-capable artillery, are arguably the most likely source of accidental use in a crisis, in particular since it is appropriate to include conventional war under the rubric of "crisis" in the nuclear age. These weapons have a very cumbersome command system to prevent unauthorized use, and for that reason it is possible that circumstances could arise in which some level of pre-delegation would be authorized. At the same time, these weapons are likely to be developed by combat. Very strong (possibly unanimous) support was given to the proposition that if the Vienna negotiations are successful, and lead to conventional parity between NATO and the WTO, the two alliances should eliminate all ground-based nuclear weapons.

2.5 Accidental War via Systematic Failure Modes of the Command System

The preceding discussion focused on the possibility that accidental (or unauthorized) launches could serve as triggers for accidental nuclear war, especially should the side subjected to such a launch respond with prompt launch. But risks that may be more grave, in part because they are less tangible, may lurk in the hardware/software systems that have become indispensable to military and political command organizations.

It is common knowledge that civilian (and especially software systems) of a similar type, when first put into operation, are plagued by bizarre malfunctions that were unforeseen and that are usually only resolved after protracted "debugging". In the case of systems that are upgraded by elaboration and conglomeration over long periods of time, instead of wholesale redesign and replacement, there is the further problem that hardly any of the system's operators understand it in its entirety. In addition, there is the problem of the ever-growing gap between the sophistication of such systems, and that of the bulk of its operators.

Both superpowers' strategic command systems rely on hardware-software systems of this character: they have been developed over decades, and because of their size must be staffed by personnel that, on the whole, are only trained to deal with a rather limited spectrum of circumstances. The sharpest contrast with civilian systems, of course, is that these politico-military systems have never operated under highly stressing circumstances – when both sides are at a high alert status (say at DEFCON 2 and its Soviet counterpart). In that case, furthermore, they would no longer be two systems, because their mutual real-time interactions would have turned them, in effect, into a single complex. In peacetime it is only possible to test the system, and its personnel, to a limited degree by simulating such conditions. There is good reason, therefore, to be concerned that under highly stressing circumstances catastrophic failure modes could rear their heads.

By their nature, one cannot foresee how such failure modes would arise, but plausible scenarios are easily constructed. For example, under severe circumstances, tactical warning signals that the system's human component would reject as false in peacetime could, instead, open paths to large-scale failure modes because of the stress on those people, and/or the far larger data flow into the command system and the more rapid decision pace that it would be expected to meet. By the same token, an accidental launch that would be recognized as such in peacetime might in a severe crisis stimulate a catastrophic response because of a systematic malfunction.

The concern that their command system might malfunction under conditions in which it has never operated is one of the driving forces behind both militaries' interest in prompt launch. This provides yet another reason why the forces, command systems, and above all political structures, must be refashioned to make any realistic "test" of this whole system as remote a contingency as possible.

Hence, the possibility of accidental war being triggered by "false warning" that enemy missiles are on the way must be taken seriously even though in the past all potential false warnings in the US system have been clarified and rejected at some stage of the warning and/or decision-making process. (High-ranking Soviet military leaders have privately acknowledged that the Soviet Union, too, has experienced "false warnings" but that Soviet authorities have increased the dependability of their warning system and the reliability of their control system).

It has to be understood that there are several steps in the process of identifying indicators of possible attack in interpreting their significance and that, therefore, the term "false warning" is often applied very loosely and somewhat misleadingly to the initial stages of the process. An empirical study that found an apparent correlation between the US policymakers' perception of increased tension in US-Soviet relations and the frequency of Threat Evaluation Conferences in the NORAD C^3I system was the subject of considerable discussion in the Workshop since it seems to lend support to concerns that "false warning" is more likely during a tense crisis. However, various questions were raised about some of the assumptions and the methodology of the study, and as to whether the correlation in question had the causal significance attributed to it. It was agreed that follow-up research on this important subject is highly desirable.

3. Inadvertent War

Inadvertent war received less discussion in the Workshop. In some usages, the term "inadvertent" refers to a scenario in which a crisis gets out of control and escalates to the point at which one or the other side (or both) come to believe that war is now virtually inevitable and seemingly imminent, and that it is better to initiate a pre-emptive nuclear strike before the opponent does so. It is an unwanted war in that at the outset of the crisis neither side wanted nor expected war but they drift and stumble into it as a result of provocative coercive bargaining, misperceptions, misjudgments and perhaps accidental or unauthorized actions, all of which feed escalation to the point at which a time-urgent decision whether to pre-empt seems to be necessary. Historical instances of war that neither side expected or wanted at the beginning of a crisis provide other scenarios of inadvertent war which are less relevant to our immediate discussion. (The emphasis here, that inadvertent war could result from a crisis that escalates and spins out of control, by no means implies that a crisis is not also one of the contexts in which an accidental war might take place).

Many of the technical issues related to accidents that could contribute to crisis escalation that results in inadvertent war are already covered by the preceding discussion of accidental war, and the measures proposed there would, to a considerable degree, ameliorate the risk of inadvertent nuclear war. But this risk is fundamentally a political one, since it depends on a decision to pre-empt made by a top-level, responsible civilian authority.

Therefore, everyone agrees that it is essential to avert conventional war for that would provide highly fertile ground for escalation leading to a possible pre-emptive strike. And by that token, it is also essential to prevent political differences from escalating into dangerous crises. The superpowers' behaviour since the Cuban crisis indicates that they have grasped this point.

Measures that would reduce the risk of crisis escalation and the outbreak of conventional war between nuclear armed states must, in consequence of the preceding observation, be largely political. The Confidence and Security Building Measures (CSBM) regime now under discussion at Vienna shows the promise of creating a truly new situation wherein the dangers inherent to the security dilemma will be substantially reduced. At the moment there is still little thought on, let alone movement towards, CSBMs that would serve the same purpose in the strategic confrontation. This is an area that warrants further work. The ballistic missile submarine stand-off zone, and a set of CSBMs that would constrain military operations in space as part of an ASAT arms control regime, are two possibilities that were already mentioned.

4. Effects of Crisis-Induced Stress on Decision Making

"That kind of crisis-induced pressure (during the Cuban missile crisis) does strange things to a human being, even to brilliant, self-confident, mature, experienced men. For some it brings out characteristics and strengths that perhaps even they never knew they had, and for others the pressure is too overwhelming". (Robert F. Kennedy, *Thirteen Days*, 1969; p. 22).

"I saw, first-hand, during the long days and nights of the Cuban crisis, how brutally physical and mental fatigue can numb the good sense as well as the senses of normally articulate men". (Theodore C. Sorensen, *Decision-Making in the White House*, 1964; p. 76).

A former high official in the Kennedy administration was asked several years ago to clarify these cryptic observations made by Robert Kennedy and Theodore Sorensen. He replied that two important members of the President's advisory group had been unable to cope with the stress of the crisis; they became quite passive and were unable to fulfil their responsibilities. Their condition was very noticeable, however, and others took over their duties. Inability to cope with crisis-induced stress has been reported for other leaders as well; Stalin, after the Nazi attack on the Soviet Union, Prime Minister Anthony Eden, at the time of the Suez crisis in 1956, Israeli Army Chief of Staff Yitzhak Rabin, during the 1967 Middle East crisis, India's Jawaharlal Nehru in 1962, and Egypt's Abdel Nasser in 1967.

In these cases the dysfunctional effects of stress on the policy maker were highly visible and could be easily recognized by others, thus providing opportunities for timely intervention and compensatory action by others. But, as will be noted, extreme stress can also have less visible and therefore more insidious effects on the performance of decision-making tasks.

A great deal is known about the effects of stress on the performance of a variety of tasks. A mild level of stress generally improves an individual's or a group's performance but as stress increases to higher levels, and/or is prolonged, performance worsens. The point at which this cross-over occurs varies for different individuals and groups, so that, fortunately, one would expect – as in the Cuban missile crisis – some members of a decision-making group to continue to function effectively even though the performance of others has sharply deteriorated.

The following is a brief summary of the major types of harmful effects of extreme stress on performance of complex cognitive tasks of the kind associated with crisis decision-making. Most of these effects are of the insidious, not the easily recognizable kind referred to above:

1. *Impaired attention and perception*: (a) important aspects of the crisis situation may escape scrutiny; (b) conflicting values and interests at stake may be overlooked; (c) the range of perceived action alternatives is likely to narrow but not necessarily to the best option; and (d) the search for relevant information and options tends to be dominated by past experience, with a tendency to fall back on familiar solutions that have worked in the past, whether or not they are appropriate to the present situation.
2. *Increased cognitive rigidity*: (a) impaired ability to improvise and reduced creativity in problem-solving; (b) reduced receptivity to information that challenges existing beliefs and preferences; (c) increased stereotype thinking; and (d) reduced tolerance of ambiguity, which results in a tendency to cut off information search and option evaluation, and to make decisions prematurely.
3. *Shortened and narrowed perspectives*: (a) less attention to longer-range consequences of options; (b) less attention to side-effects of options.
4. *Shifting the burden to the opponent*: (a) belief that one's own options for dealing with the crisis and avoiding escalation are quite limited; and (b) belief that only the opponent has it within his power to prevent an impending disaster.

The emotional climate of crisis decision-making (anger, fear, distrust) exacerbates these tendencies, contributes to misperceptions and misjudgments and may provide impulses for immediate strong action of some kind. Under certain conditions decision-making groups that enjoy strong group cohesion may attempt to cope with the cognitive and moral stresses experienced in a crisis by engaging in concurrence-seeking, avoiding challenges to each other, and providing mutual support, thus falling victim to the phenomenon of "group-think".

Responsibility for shielding the decision-making process from the adverse effects of stress on some members of the group must be shared by its members. There must be a willingness and a capacity to recognize when some individual, or the group as a whole, is no longer able to handle the psychological stress, the emotional and physical fatigue of a tense crisis. In addition, methods and techniques for stress avoidance and stress management need to be incorporated into the procedures of the policy-making system and, indeed, into the medical support system available to top officials.

There is almost universal consensus among specialists on decision-making as to *standards and practices* that a decision-making group should adhere to in processing information, generating alternative courses of action, and evaluating options before decisions are made. These standards are well known and need not be repeated here. Suffice to say that adherence to such good decision-making practices will itself help to reduce possibly dysfunctional effects of stress. But for such benefits to be realized these standards must be embedded in the norms of the policy-making group; they must be internalized by all participants; and the decision-making process in a crisis must be carefully monitored by the executive and/or a surrogate charged with this responsibility (i.e. the "custodian-manager" of the process) in order to spot possibly serious deviations from these norms and practices, and to take corrective action on a timely basis.

In addition, the amount and quality of *preparatory work* (i.e., anticipating and analysing a variety of hypothetical crises) has been shown to be important in reducing the dysfunctional effects of stress. While crisis simulations on a classified basis are known to take place at a relatively high level in the US government (and perhaps in the Soviet government), participation seldom extends to the top-level leaders who will be called upon to make the critically important decisions in real-life crises. And the focus of such crisis simulations seems to be directed more to improvement of policy planning rather than also to generating the kinds of preparatory work that could reduce vulnerability to stress or increase familiarity with and practice in applying crisis management precepts.

5. Effects of Organic Illness and Sleep Deprivation on Decision Making

The possibility that major psychiatric impairment in a key official may not be detected or controlled in time to prevent a policy disaster, and that crisis-induced stress may activate latent psychological vulnerabilities or psychotic tendencies in an individual, are matters of legitimate and grave concern. However, these possible sources of accidental or inadvertent nuclear war were not singled out for discussion in our meeting.

Some attention, however, was given to the effects that various organic illnesses, emotional and physical fatigue, and sleep deprivation can have on crisis decision-making. Medical problems such as heart disease, strokes, traumas, surgery, cancer, hypertension, blood loss are known to produce deficits in the afflicted individual's performance of various cognitive functions – such as the ability to concentrate, the span of attention, memory, inventiveness, deductive and inductive reasoning. A manifest overlap exists between cognitive effects of organic disease and those of extreme stress that were alluded to above. Illness can heighten vulnerability to crisis-induced stress. Moreover, insufficient attention has been given to the likelihood that the side effects of many medications can contribute to impairment of cognitive functioning.

It is of critical importance to establish and make timely use of procedures for identifying temporarily disabled heads of state and other key officials and removing them from their primary decision-making role. Clarification and rehearsal of such procedures should occur from the first days of a new administration and be repeated periodically.

Pronounced sleep deprivation and disruption of an individual's circadian rhythms (i.e., the "biological clock") can also result in impaired cognitive functioning. This problem can arise also with odd sleep cycles that are in conflict with a human being's natural circadian system, possibly resulting in insomnia, emotional disturbance, impaired information processing and judgment. (It has been reported that pronounced sleep deprivation of key individuals contributed to the faulty decision-making process that led to the Challenger accident).

After a day or two of sleep deprivation, stimulant medication, such as amphetamines, has been shown to reverse most of the documented performance impairment. This might be useful in a short-term situation, but the harmful potential of prolonged use of stimulants by crisis decision-makers and other key personnel in the chain of command would be especially troublesome in the nuclear era, for it risks inducing a paranoid state of mind accompanied by greatly heightened impulsiveness and very poor judgment.

The medical support system available to top policy-makers in the past has functioned erratically and at times inadequately. This system should be periodically reviewed and upgraded as necessary to provide comprehensive quality care not only for the treatment of organic illness and careful monitoring for possibly adverse side-effects of medications; in addition, it should provide effective monitoring and advice with regard to dysfunctional effects of extreme stress, avoidance of debilitating fatigue from overwork and inadequate sleep, and appropriate counselling with respect to the adverse effects of over reliance on sleeping pills and amphetamines.

6. Crisis Management and Crisis Avoidance

As mentioned earlier, maintaining the quality of the decision-making process during a crisis is one way of providing safeguards against the harmful impact of psychological stress, organic illness, sleep deprivation, emotional and physical fatigue. Another way is to ensure that participants in the decision-making group are thoroughly familiar with the general principles and requirements of crisis management and have acquired appropriate training to equip them to deal with the challenging task of applying these general principles to the special characteristics of a particular crisis that they are forced to deal with.

To be sure, all past superpower crises were managed and terminated without any kind of shooting war between US and Soviet forces and, indeed, there has not been a war-threatening superpower confrontation since the October 1973 war in the Middle East. Nonetheless, we believe that the possibility of additional superpower crises of a war-threatening nature cannot be ignored, and that there is no guarantee that crisis management efforts would be equally successful in the future. In fact, developments of the past decade in military systems and deployment have raised serious questions among informed observers as to whether crisis management concepts and procedures would be robust enough to withstand the escalatory pressures inherent in any new superpower confrontation.

Given the widely shared view that if nuclear war occurs – whether "accidental" or "inadvertent" – it is likely to do so during the course of a tense superpower confrontation, the highest priority must be given to continuing efforts to avoid war-threatening crises and to strengthening crisis management capabilities. In the limited space available only a few suggestions can be offered.

1. When conflicts of interest are simmering, threatening to erupt into a dangerous crisis, the superpowers should rely in the first instance on timely diplomatic communication and negotiation rather than engage in efforts to protect their interests by unilateral actions that are likely to provoke counter moves and crisis escalation.

2. In peacetime, as well as when embroiled in a dispute, the superpowers should avoid sudden and/or secret military deployments that affect the strategic or regional military balance; and they should forgo foreign policy initiatives that seek to make major gains at each other's expense. Unless American and Soviet leaders observe such restraints in their behaviour towards each other, high-sounding declarations that they will forgo use of force and threats of force against each other will remain hollow and brittle.

3. Control and restraint of one's military forces is a critical requirement for effective crisis management. Top-level leaders on both sides must exercise informed and judicious control over all movements and alerts of combat forces. Leaders must be sensitive to and deal constructively with the tension that frequently exists in crisis situations between the imperatives of "military logic" and the requirements of diplomacy if they are to prevent unwanted escalation in crisis situations. Similarly, it is of the utmost importance that the rules of engagement given to combat forces be carefully chosen and monitored to reduce and control the risks of unwanted escalation.

4. When a war-threatening crisis erupts, each side may find it necessary to take measures to reduce the vulnerability of its theatre forces and to enhance their readiness to *defend* themselves. However, at the same time, American and Soviet leaders should take the greatest care to avoid alerts and deployments that are likely to be perceived by the other side as preparations for significant and imminent *offensive* operations. They should also avoid (or at least greatly mini-

mize) the practice of using military alerts and deployments to signal "resolution", and/or to exert coercive pressure on the adversary for crisis bargaining purposes. In brief, there should be more use of diplomatic communication and less reliance on military "signalling" *per se*.

5. The task of crisis management should not focus exclusively on avoiding an accidental or inadvertent nuclear war only in the circumstances when a crisis has already escalated to high-level alerts. However important it is to increase the reliability of warning systems, to eliminate the possibility of launch on false warning, to avoid or neutralize the effect of accidental or unauthorized actions and human errors, etc., elementary prudence demands that maximal efforts be made to prevent a conflict of interest from developing into a tense US-Soviet crisis which will severely test their ability to avoid accidental or inadvertent war. Accordingly, the *first* objective of crisis management should be to terminate any US-Soviet crisis promptly and effectively at its outset before it is allowed to escalate to dangerously higher levels. US and Soviet leaders should discuss with each other in peace-time ways of preventing crisis escalation in different contexts, including regional conflicts in which they are supporting rival regional actors.

6. The time factor may be critical if efforts to avoid escalation and terminate a crisis are to have an opportunity to succeed. Accordingly, leaders on both sides must recognize the necessity to slow down the tempo and momentum of crisis developments. They may need to deliberately create pauses in crisis activity – as, for example, President Kennedy did in the Cuban missile crisis – in order to provide enough time for the two sides to exchange diplomatic communications and to give the leadership on both sides adequate time to assess the situation, make well-considered decisions, respond to each other's proposals, etc.

7. Limitation of the objectives one pursues in a confrontation with the other superpower and resisting the temptation to inflict a damaging, humiliating defeat on the opponent may be essential if a superpower crisis is to be settled without escalation to dangerously high alert levels. While seeking to protect its own fundamental interests, each side must recognize the other side's legitimate interests and strive for a mutually acceptable formula for terminating the crisis.

8. A great deal of essential knowledge and relevant experience has been gained from successfully managing past US-Soviet crises. But the learning experience is virtually worthless unless the lessons and requisite skills are codified and institutionalized in each government. This body of knowledge and know-how must be transmitted quickly and effectively to new leaders and staff members who will have important crisis management responsibilities. It must be internalized by them via appropriate training, exercises and rehearsals so that they will perform effectively the very first time they are called upon to manage a crisis.

9. Finally, since effective control and management of US-Soviet confrontations may depend critically on the ability of the two sides to avoid misperceptions and miscalculations and to communicate rapidly and effectively with each other, it is highly desirable for Soviet and American military and civilian specialists to

discuss together problems of crisis stability, to identify types of behaviours that may threaten loss of control and trigger escalation pressures, and to consider and make provisions for preventive and remedial measures.

7. Managing and Improving the Overall US-Soviet Relationship

Members of the Workshop emphasized that the state of overall US-Soviet political relations is a contractual factor that influences whether accidental and unauthorized military actions and events stemming from human error and frailty will contribute to processes that lead to accidental or inadvertent war. Progress in arms control cannot proceed expeditiously and in full measure except in conjunction with improvement in the overall relationship.

During the past forty years US and Soviet leaders have developed and generally adhered to a number of fundamental though largely tacit "*rules of prudence*", and this has helped to limit the conflict potential of their relationship, even during its most tense periods. These tacit "rules" should be discussed, clarified, elaborated and strengthened. They include, but are not limited to the following. (For additional discussion see G. T. Allison, W. L. Ury and B. J. Allyn (editors), *Windows of Opportunity: From Cold War to Peaceful Competition in US-Soviet Relations* (Cambridge, MA, 1989); Chapter 1, "Primitive Rules of Prudence", by Graham T. Allison).

1. Respect and do not challenge each other's "vital interests"; this "rule" has an important corollary; avoid provocative efforts to exploit difficulties the other superpower is having in its sphere of influence or in its efforts to withdraw from advanced positions in the Third World.

2. Do not initiate military action of any kind against the other superpower's military forces. Whatever the likelihood that even low-level conventional military clashes could escalate to nuclear war, the risk is not acceptable. This "rule of prudence" was recognized and endorsed in the joint communique issued by Reagan and Gorbachev at their first summit in 1985: "Recognizing that any conflict between the USSR and the USA could have catastrophic consequences, they emphasized the importance of preventing any war between them, whether nuclear or conventional".

3. In superpower crises (such as in Berlin, Cuba, the Middle East in the past) do not exploit an advantage to the point that the other side is forced to make the awesome choice between accepting a humiliating diplomatic defeat or initiating use of military force in order to prevent such a setback.

4. When the two superpowers are backing rival regional actors in a Third Area conflict (e.g., the Middle East), each superpower should accept responsibility for preventing its regional ally from inflicting a shattering defeat on the regional ally of the other superpower, thereby placing that superpower in the position

of having to intervene. If a superpower is unable or unwilling to discharge its responsibility to restrain its victorious ally, it should accept the necessity for the other superpower's intervention as a last resort to prevent its regional ally from suffering such a defeat.

These tacit "rules of prudence" need to be supplemented by more positive "*rules of cooperation*" in the interest of developing a more constructive and stable bilateral relationship and a safer and more viable global society. The superpowers should seek to reach agreements that will enable them to deal more effectively with predictably sensitive and potentially explosive situations that affect not only their own well-being but that of other peoples as well. They should strengthen existing procedures and create new institutional mechanisms to address the variety of issues that already are or could become highly dangerous to peace, that are corrosive of their overall relationship or harmful to the global community. They should increase efforts to develop better understanding and communication between political and military leaders and officials at all levels in the interest of creating a *sense of partnership* for dealing with these threats.

The Soviet Union and the United States have already made substantial progress in developing a variety of cooperative arrangements for enhancing their security. Leaders on both sides have gradually come to understand that they can no longer assure the security of their own country at acceptable cost and risk through strictly unilateral national efforts as in the past. The realization that each superpower is dependent for its security on the behaviour of the other – i.e., a condition of mutual vulnerability and mutual dependence – has provided powerful incentives for security cooperation. Both superpowers are moving towards a realization that the traditional conception of "national security" must be broadened to include a new conception of "mutual security".

Under changing conditions that create new challenges and offer new opportunities, *theories of cooperation and conflict resolution* formulated by psychologists, game theorists and other specialists assume enhanced practical import. Foreign policy officials in both countries should acquire thorough familiarity with these theories and should consider how they can be adapted and utilized in efforts to develop a more constructive and stable US-Soviet relationship and a more viable and livable global society.

Part II

European Security

The Vienna Force Reductions Talks: Resolved and Unresolved Issues

Jonathan Dean

1. Introduction

The prospects are quite good for step-by-step build-down of the East-West military confrontation in Europe through a combination of negotiated and unilateral measures. Given timely action by NATO, a first agreement in the Vienna Conventional Armed Forces reduction talks along the lines of NATO proposals is feasible – even probable – within the next two to three years. A partial agreement could be concluded earlier. Only major negative developments, in particular, the recurrence of East-West and West-West disputes over nuclear weapons, or the replacement of the Gorbachev leadership by a less constructive one, seem likely to be able to block this positive outcome.

Many specific problems remain to be dealt with at Vienna. They include: defining a mutually acceptable reduction base for combat aircraft, helicopters and military manpower; dividing the Atlantic-to-Urals area into sub-zones; agreeing on stabilizing measures and constraints; the issue of armaments stored by the United States in central Europe; and the problem of verification, especially difficult when each alliance will be allowed millions of personnel and over 100 000 treaty regulated items, as compared with the 1600 delivery systems to be checked in a START treaty. The scope and rapidity of initial progress at the Conventional Forces in Europe (CFE) indicates that solutions will be found for these problems.

In this paper, I will review what has been accomplished and what remains to be done to reach a first agreement and to place the negotiating process firmly on the tracks of continuing build-down, including action to deal with the nuclear issue.

2. Present Status

In the first three months of the Vienna talks, there was rapid movement, first by the Warsaw Pact (WTO) and then by NATO. The two alliances now agree on the force components to be reduced: tanks, artillery armoured troop carriers, combat aircraft, armed helicopters, active-duty ground and air force personnel and, in a later negotiation, tactical nuclear armaments. And they also agree on a reduction method which involves a percentage cut from the current holdings of the numerically weaker side, NATO in most of these cases, with the Warsaw Pact coming down to the new NATO levels. In several cases, they have agreed on the actual numerical levels themselves.

This progress in resolving the main issues of principle in the talks is startling. It places the participants in the CFE talks after only three months of negotiation about where the USA and USSR were in the INF negotiations after four years of negotiation.

In the short period between 9 March 1989, when the CFE talks began, and the end of May, Warsaw Pact negotiators indicated willingness in principle to accept five main elements of NATO's highly demanding reduction proposals. These are:

1. Reducing the heavy main battle tanks of both alliances to 20 000. (Using NATO data, this would mean a WTO reduction of about 30 000 tanks and a NATO reduction of 2000).
2. Reducing armoured troop carriers to an equal level of 28 000 per alliance. (This would mean a WTO reduction of about 42 000 and a NATO reduction of perhaps under 1000).
3. Reducing artillery to a common ceiling of 16 500. (This would mean a WTO reduction of 27 000 artillery pieces and a NATO reduction of only about 1200. There are definitional differences on artillery and also in the other two armaments, but they are soluble).
4. Placing a sub-ceiling on Soviet holdings of remaining armaments of the type reduced at 60 per cent of the overall total for the Warsaw Pact; for example, 60 per cent of the 20 000 tank limit for WTO or a limit of 12 000 tanks; this would be a national ceiling on Soviet forces west of the Urals which no other participant would have – because none have such large forces.
5. Placing a "stationed forces" sub-ceiling on Soviet holdings of tanks, artillery and armoured troop carriers deployed in Eastern Europe. This sub-ceiling would be equal to the combined holdings of these armaments of US, UK, French, Netherlands, Belgian and Canadian forces stationed on the territory of other allies, mainly the Federal Republic of Germany. NATO has proposed that this ceiling be set at 3200 for tanks, 6000 for armoured troop carriers and 1700 for artillery.

The Soviet Union has not yet agreed to the specific stationed forces levels suggested by NATO; there is disagreement over 6000 stored NATO tanks, mainly United States tanks. But Soviet agreement in principle to the idea of such limits means that a CFE agreement would contain a ceiling for Soviet forces stationed in Eastern Europe which would be an important obstacle to forward movement of Soviet forces to build up for attack or to reinforce political control – an obstacle which might be decisive in the light of already high Soviet reluctance to intervene in Eastern Europe.

Against the background of the unilateral reductions announced by Gorbachev in December 1988, these far-reaching Soviet moves at Vienna are unambiguous evidence of a Soviet desire to build down the East-West confrontation in Europe. If the Soviet leadership had been interested only in public relations impact in the West, it could have effectively publicized the current implementation of the Soviet unilateral reductions and sat it out at the negotiating table at Vienna, leaving it to NATO to make the first move.

Let us now look at President Bush's proposals at NATO's fortieth anniversary summit meeting (29–30 May 1989) in Brussels of the heads of government of member states. In essence, the Bush proposals are as follows:

1. To include land-based combat aircraft and helicopters in negotiated reductions and to cut them to an equal level set at 15 per cent below current NATO levels.
2. To reduce United States and Soviet ground and air force personnel to an equal level of 275 000 men in the Central European sub-region.
3. To accelerate the timetable for reaching a first NATO-Pact agreement on force reductions in order to achieve agreement within six to twelve months from September, 1989.
4. There is a fourth point, that reduced armaments be destroyed. But that point had already been advanced in the negotiations to meet an earlier Soviet proposal. It is the right principle, but will not prevent both alliances from destroying their older equipment first.

3. Aircraft Reductions

President Bush's proposal to reduce combat aircraft and helicopters is the most important of these moves. Since the outset of CFE, the Soviets have insisted that combat aircraft be reduced as part of an overall agreement eliminating the Warsaw Pact's large superiorities in tanks, artillery and armoured troop carriers. It is likely that the Pact would have made inclusion of aircraft reductions a requirement for any CFE agreement. Moreover, given the agreed aim of the new Vienna talks to reduce the capacity of either side to attack, it is wholly logical to include in reductions aircraft and helicopters, which are essential to the success of any offensive operation.

President Bush's aircraft reduction proposal has eliminated an estimated two years of debate within NATO about when to move on this subject, and has thus brought closer an early East-West agreement. The President's suggestion that the aircraft cut be at 15 per cent of current levels generally corresponds to the dimensions of the cut proposed by WTO and should be acceptable as regards the size of the cut.

But a 15 per cent cut of what? As yet, there is no East-West agreement on which aircraft will be reduced by both sides. The Soviet Union, which has counted ground-attack aircraft and fighter-bombers on both sides, but excluded everything else, appears to be claiming that NATO has about 4000 aircraft and WTO about 2800. In a 15 per cent reduction from the current level of the weaker side, WTO would reduce about 400 aircraft to a level of 2400. NATO would reduce 1600 aircraft.

For its part, NATO is counting in its data all fighter interceptors and training aircraft, as well as ground-attack aircraft, fighter-bombers and medium bombers, for a WTO total of roughly 14 000 aircraft to NATO's 6600 (including "combat-capable" trainers). Using this reduction base, after a 15 per cent cut in the NATO level to a common ceiling of 5700, NATO would reduce about 900 aircraft and the WTO about 8000.

That is a lot of airplanes, even in a situation where the Soviets have already indicated willingness to make much larger reductions of ground force armaments than NATO; this looks like an issue which will have to be resolved by a high-level political compromise. In my own approach to establishing a reduction base for aircraft, I include ground-attack aircraft, fighter-bombers, medium bombers and those air-to-air interceptor aircraft which are multi-purpose, like the American F-16 and Soviet Mig-21, and which can be used both for ground attack and air-to-air interception. I exclude training and support aircraft like AWACS, land-based naval aircraft – NATO is excluding its own carrier-based aircraft because they belong to naval forces, which are excluded from coverage in CFE by the agreed mandate – and the remaining interceptors. A 15 per cent reduction using this modified reduction base would result in a reduction of about 700 NATO aircraft and 550 WTO aircraft. (See Appendix).

4. Manpower Reductions

President Bush's proposal that United States and Soviet manpower in Central Europe be reduced, is valuable because it cuts through NATO's previous reluctance to include military personnel in reductions. This reluctance is based on the difficulty of verification and the negative data experience in MBFR. The Bush proposal partially meets the WTO proposal that negotiated reductions should cover military personnel as well as armaments. The proposal foresees a 10 per cent reduction of about 30 000 US personnel deployed in Central Europe, as compared with a reduction of 300 000 forward-deployed Soviet personnel down to an equal level of 270 000 men. The Soviets have proposed a stationed forces limit of 375 000 in Central Europe. But they included in the Western quota the more than 100 000 personnel of French, British, Canadian, Netherlands and Belgian units stationed in West Germany.

The Western proposal to reduce only US manpower for NATO is inconsistent with the fact that NATO has included the armaments of allied forces stationed in West Germany (those listed above) in its proposals for equal levels in the armaments held by stationed forces of both alliances deployed in Central Europe.

Moreover, the Bush proposal does not cover Soviet military personnel deployed in Soviet territory west of the Urals. These personnel should be reduced or limited to give NATO some control over Soviet mobilization potential, and to preclude circumvention through increasing the personnel of permitted units. And, without negotiated manpower reductions for all NATO countries, it will be very difficult for NATO countries other than the United States to realize savings from force cuts.

My own proposal for reducing manpower in a first-phase agreement, is to cut to an equal ceiling at 10 per cent of the NATO level the active-duty ground and air force personnel of all CFE participants in the Central European area proposed by the Warsaw Pact (Denmark, Benelux, the Federal Republic of Germany on the West; and the GDR, Czechoslovakia, Poland and Hungary on the East – a more logical Central European Zone than NATO'S, which omits Hungary and Denmark). Under that equal ceiling, there would be a sub-ceiling for stationed force personnel, like that proposed by NATO covering US and Soviet forces, or Soviet forces on one side and all allied foreign forces on the other. Ground and air force personnel of both alliances in the Atlantic-to-Urals area outside of Central Europe would be placed under a no-increase ceiling and reduced in a second phase of negotiation.

Time will be needed to resolve these difficulties. It will be very difficult to achieve an agreement complete in all details within the six- to twelve-month period President Bush has suggested – a two- to three-year timetable seems more realistic. However, President Bush's proposal for agreement within six to twelve months represents his personal engagement in favour of an early agreement in the CFE talks. The President's commitment evidences his administration's belief that the Gorbachev policy does represent a genuine opportunity to make deep cuts in Soviet forces. The President will have to continue to devote considerable time to CFE amid many pressing requirements, if he is to provide the impetus necessary for an early outcome. To deal with the issue, he should appoint a Washington-based presidential-level coordinator for the CFE talks.

5. Some Shortcomings in the NATO and WTO Positions

NATO's current position has some faults of design relevant for the long term. Some are also found in the Warsaw Pact position. I see six of these. Work is underway in NATO on some of these, but nothing is being done on others.

1. The NATO approach to CFE as yet contains no specific proposals for early-warning measures. These are an essential aspect of a CFE agreement, especially if our aim is, as it should be, to design agreements which maximize insurance against recurrence of a more expansionist Soviet policy, and if the Warsaw Treaty countries want similar assurance as regards NATO.

I suggest that we should make the entire Atlantic-to-Urals area a Zone of Military Transparency. One component of this zone would be a reciprocal quota of 150 overflights a year by low-flying aircraft over each half of the Atlantic-to-Urals area, going all the way back into the Soviet Union to cover reserve and mobilization units. This suggestion would be a useful application of President Bush's Open Skies proposal, and it could be combined with Warsaw Pact proposals for a limited quota of exit-entry points and observers at military headquarters, major airfields and major traffic choke-points, or with teams of roving observers with a similar function.

2. The NATO approach does not now provide for constraints on forward deployment of especially threatening armaments, ammunition stocks or equipment. Neither the Central European sub-region nor the restriction on equipment of forces stationed on foreign territory which NATO proposes would prevent such forward deployments.

If we are going to limit forward deployment of one item, then it might as well be done for several. That is why I argue for an asymmetrical Restricted Military Area 100 kilometers deep on the WTO side, and fifty kilometers on the NATO side, from which reducible armaments (tanks, artillery, armoured troop carriers and combat aircraft and helicopters), large ammunition stocks, tank transporters, pipe-laying equipment and bridging would be eliminated or thinned out. A zone like this would in practice move NATO's defensive frontier 100 kilometers to the east. Such zones could be repeated elsewhere in the north and in the south. They would be a way of compelling defensive restructuring to the forces of both alliances at their point of contact. The Warsaw Pact has made a proposal for symmetrical thin-out zones of this type.

3. The NATO proposal contains no provision for ceilings on units of the type reduced. WTO and NATO could reestablish reduced tank units, or even increase their number, using a smaller number of tanks, but a larger number of supporting arms and infantry. NATO commanders desire flexibility on this issue. But the present position does not seem right if the aim is to bring about enduring reduction of military capability on both sides.

4. Other than the proposal for reducing US and Soviet personnel in Central Europe to an equal level, the NATO proposals contain no provision for manpower ceilings.

This omission still rests on NATO's belief that manpower ceilings would be difficult to verify. But it would permit the Soviet Union to increase without limit the manpower of combat units deployed in the USSR, including reserve units, even if it wished to reduce overall military manpower for economic reasons. NATO needs manpower ceilings to control the Soviet reserves. The Warsaw Pact also has a legitimate interest in controlling the size of Western forces and reserves. Earlier in this paper, I have made a suggestion for a phase-one manpower cut.

5. Neither NATO nor WTO approaches contain provision for controlling the continuing production in the reduction area of armaments reduced by agreement. Soviet tank and artillery production would continue unrestricted. The same goes for that of NATO countries.

6. Surprisingly, given the new openness of the Soviets, the NATO approach as it now stands does not provide for full and detailed data exchange early in the CFE talks. Data exchange and data validation inspections are essential in order to begin the verification process from the outset of the talks, much as the Bush administration is proposing for the START talks.

Taken together, these shortcomings detract considerably from the military value of the NATO proposal and, if uncorrected, would make it easier for either side to circumvent it in the future.

The current NATO approach has a further major defect. It would do little to reduce the costs of the NATO-Pact confrontation. A 5-10 per cent armament reduction would produce very little savings in the defence budgets of the NATO allies. We have to make deeper cuts than that to achieve savings. Moreover, personnel have to be cut across-the-board to reduce costs. They make up 50–60 per cent of NATO defence budgets.

NATO should not change the dimensions of its first-phase proposal, because it appears feasible to achieve an agreement along these lines fairly soon. But it should broaden the scope of its first-stage proposal to include early-warning measures, constraints, and ceilings on reduced units and on overall military manpower.

6. Deep Cuts

NATO should expand its overall reduction objective and develop a specific proposal for the second phase of cuts to which both alliances have committed themselves in general terms.

For the past several years, I have been arguing for an overall programme of moving to a lower equal plateau between the alliances, based on a 50 per cent cut in NATO's current holdings of the offensive armaments which have now been selected for reduction – that is, tanks, artillery, troop carriers, combat aircraft, helicopters and, in the future, surface-to-surface missiles, as well as a 50 per cent cut from the present level of NATO active-duty ground and air force personnel. Other, more defensive armaments would not be cut and their relative strengths in the reduced forces of both alliances would increase.

This programme would be carried out over a period of ten years and should result in cuts of about 30 per cent in the budgets of NATO countries for European defence. In practical terms, this means cutting another 40 per cent of NATO reducible armaments after a first phase, adding missiles and overall manpower to reductions, and bringing WTO down to these low levels. The latter has said it is ready to go down an additional 25 per cent from the equal levels reached in phase one in a second phase. It should go still further.

Looking at the Central European Zone proposed by WTO – as noted, this seems preferable to NATO's Central European as a sub-region because it includes Hungary and Denmark – the reduction I propose would leave each alliance about 4000 tanks, 2000 artillery pieces and 5000 troop carriers in central Europe. About half of WTO's quota could be Soviet, enough equipment to provide for about ten light tank divisions or seven traditional-strength tank divisions, as compared to the 30 divisions which have been deployed there up to this year. In addition to these cuts in Central Europe, a Restricted Military Area would be established from which all armaments of the type reduced and major ammunition storage sites would be prohibited, but where each alliance would be permitted to deploy infantry, combat engineers, anti-tank weapons, scatter and pipe mines, rapidly emplaceable obstacles and other counter-mobility measures.

Taken together, these measures would create a situation where defensive armaments and troop dispositions can bring to bear greater fire-power than attacking force. Successful attack by either alliances using forward-deployed forces in place would seem excluded.

7. Dealing with Reserves and Mobilization

There remains the issue of force mobilization and force build-up, which both alliances are interested in preventing. In this regard, I do not think that NATO should deal with the existing uncertainties about the Soviet future by waiting to see what the final outcome is before making deep cuts of its own forces. Instead, the Western countries should make maximum use of the considerable flexibility being shown by the Gorbachev leadership, to cut back and restructure Soviet forces in such a way as to interpose long periods of time before a less conciliatory successor could build up these forces sufficiently to be confident about using them for aggression or intimidation.

Both alliances have agreed to eliminate the capacity for sustained large-scale offensives. The following measures could contribute to this objective:

1. The systems of early warning and restrictions on forward deployment described earlier.
2. As proposed above, further deep cuts in holdings of armaments more needed for offense than defence – those armaments already agreed for reduction. NATO's present proposal for reduction of these arms is limited to about 10 per cent of its current strength. These reductions would leave both alliances at a high equal level. This outcome would be a great improvement over the present situation of WTO superiority. But with both alliances still well

equipped with armaments suited for attack, it would not be a maximum contribution to crisis stability. Further cuts should be made in these armaments to move towards defence dominance.

3. Reduction through disbanding units and discharging active-duty manpower and placing residual ceilings on both. Also, placing controls over mobilization and reserve potential by restricting the active-duty manpower, equipment and level of training of reserve units on both sides.
4. Preventing diversion of continuing production in the Atlantic-to-Urals area of major reducible armaments to field units through placing that production in secured storage, open to inspection by the opposing alliance. In later phases of reduction, agreed ways should be found of reducing the level of production.

It would help in this regard if both alliances agreed in current discussions of military doctrine that the objective of defence in the event of conflict should be that now followed by NATO: to push the aggressor off home territory and to reestablish the *status quo* before attack. We live in an age where both alliances have nuclear arms which might well be used by either if it were facing defeat in a conventional conflict. It is advisable in such a situation that each alliance relinquish the objective of having conventional forces large enough to retaliate against an aggressor and punish him for his aggression through conclusive victory over his entire conventional forces. Agreement on the principle that counterattack capability should be limited to that needed to restore the *status quo ante* would provide a useful guideline for the type of force to be maintained by both alliances after deep cuts.

8. The Nuclear Issue

It is widely recognized that there is a direct relationship between nuclear and conventional deployments. If deep cuts in conventional forces take place, there will have to be – and there should be – cuts in tactical-range nuclear armaments. Moreover, it would be advantageous for the NATO countries if the large Warsaw Pact numerical superiority in nuclear armaments, especially in surface-to-surface missiles, were eliminated.

Yet, there is a fundamental conflict between the announced aim of the Soviet Union to eliminate these weapons in Europe through negotiation and the conviction of many Western military commanders and defence officials that retention of some of these armaments is essential for Western defence. These Western experts fear that any East-West negotiation on the subject of these armaments will lead to their elimination through the combined pressures of the Warsaw Pact negotiation position and of Western public opinion. Many of them believe that only the deployment of nuclear weapons has kept the peace in Europe for the past

40 years. These officials correctly point out that, even after deep cuts in Soviet conventional forces and in strategic nuclear forces, the Soviet Union will remain a major military and nuclear power. They consider that it will be necessary to maintain a counter-balancing Western nuclear force. Their insistence on doing so is intensified by the uncertainties over the future of the Soviet system already described. Their views are very firmly held.

There has been no war in Europe for the past four decades. I suspect that there also would have been no war if nuclear weapons had not been deployed by both sides. But no one can know this. This is one of the factors which gives the debate over nuclear weapons in Europe so much abstract intensity. But we do know with certainty that deployment of these weapons creates great caution with regard to possible conflict; that caution has been useful and will continue so.

There are also many in the West, as well as the East, who want to abolish all nuclear weapons. We can share their antipathy to these weapons. But, as a matter of practical politics, there is little prospect that this view will become agreed NATO policy in the foreseeable future. However, in the NATO countries, the two schools of opinion, that nuclear deployments continue essential, and that nuclear weapons must be limited, appear closely balanced in strength. The abolitionists are numerically superior; the adherents of deterrence more influential on government policy. Yet, those favouring elimination of tactical nuclear armaments for Europe are right that with the INF treaty and the growth of anti-nuclear feeling in Europe, the traditional versions of NATO's flexible response strategy are no longer viable.

If they fail to reach some agreement as to a future nuclear deterrent, these two bodies of Western opinion have the power to bring about deadlock which can slow down or block the possibility of negotiated cuts, both conventional and nuclear. An example is the situation in NATO in the first five months of 1989, when NATO countries sought in vain to agree on the timing of US-Soviet talks on reducing tactical-range missiles. But, as an alternative to deadlock, the two groups might be able to agree on the desirability of negotiated reductions to a lower level equal with WTO, moving towards a residual minimum deterrent of sub-strategic armaments on both sides, leaving to the future the question of whether it will be desirable and possible to eliminate these weapons entirely. This is the compromise we should seek.

The Soviet Union has played an ambivalent role in this controversy. Mainly, it has pressed publicly for acceptance of Gorbachev's policy of eliminating all nuclear armaments from Europe. But Soviet officials have also indicated readiness to negotiate for a low equal level as an interim measure, indefinitely postponing their goal of elimination. In Gorbachev's visit to France in early June 1989, he touched on both themes: he offered to increase the number of unilaterally withdrawn Soviet nuclear armaments beyond the 500 already announced if the West agreed to early negotiation on elimination, and he offered to discuss a minimum deterrent.

The Bush administration should take the initiative to discuss this issue with the major groups involved, starting with the European nuclear powers, France and the UK. To mitigate the apprehensions of these countries over long-term pressures to reduce their own nuclear assets, the United States should declare its readiness to join with them in a cooperative arrangement to assure a continuing European-based nuclear deterrent vis-a-vis the Soviet Union. The United States should also express its willingness to join with these countries in insisting on the exclusion of their nuclear-armed submarines, nuclear-capable aircraft and, in the case of France, tactical surface-to-surface missiles, from reduction in a US-Soviet negotiation on tactical nuclear armaments. As a demonstration of serious interest, it could offer to collaborate with France and the UK in developing a stand-off missile for aircraft which would have a range below the 275-mile limit in the INF treaty. These actions should lessen the resistance of both countries to early negotiated reductions of conventional forces and, when the time comes, to negotiated reduction of Soviet and United States nuclear tactical-range armaments in Europe.

The US administration should explain to political parties and groups in Europe which advocate elimination of tactical-range nuclear armaments, that their real choice for the short term is between negotiated reduction of these armaments to a lower level which will do away with the Soviet and Pact numerical superiority and place the remaining armaments on both sides under verified ceilings – and possible breakdown of the reduction process, with blockage of negotiated conventional and nuclear reductions and destructive friction within and among NATO countries - and possibly in eastern Europe as well – which could bar the road to an orderly continuing build-down of the East-West military confrontation in Europe. For many, negotiated reduction with postponement of the goal of elimination should be preferable. It may, in addition, be possible to agree with some of these groups, like the opposition Social Democrats in Germany, on the components of a NATO residual nuclear deterrent which they could accept as a transitional arrangement of unspecified duration.

The US administration should then discuss with the Soviet leadership the issue of reducing tactical nuclear armaments, setting forth its plans for a continuing but limited nuclear deterrent force in Europe, and seeking Soviet agreement to a mandate for negotiation on reducing United States and Soviet air- and land-based tactical-range nuclear armaments to a low equal level.

Such negotiations will not be easy even if both sides show flexibility. Because of ease of transport, limits may have to be world-wide, as in INF. Verifying residual levels of surface-to-surface missiles, nuclear-capable artillery shells, nuclear gravity bombs, and stand-off missiles, would present very serious problems, perhaps more difficult than those of verifying destruction of existing stocks of chemical armaments, because we are not talking now of cessation of production of nuclear arms. Even supervised destruction of reduced warheads and tight checks on continuing production would not be a full answer. In the light of these difficulties, partial step-by-step measures of reduction and of verification may be needed. Despite these uncertainties, as far as the NATO countries are concerned, the elimination of at least the current large WTO superiority in surface-to-surface missile launchers, given the fact that each side would continue to have its own tactical and strategic deterrent to possible use of nuclear weapons, would be an advance over the present situation.

Unless more encompassing restrictions on modernization can be agreed for both conventional and nuclear armaments reduced by negotiation, it should be permissible to modernize remaining nuclear armaments as long as their range remains below the 500-kilometer limit established by the INF treaty. It would be illogical to come to a difficult compromise on nuclear arms and then to find that obsolescent Lance missiles could not be replaced or that improved stand-off missiles could not be deployed. It is particularly urgent to find some solution for standoff missiles, or West-West and East-West disputes over them may become more than in the case of Lance.

The UK administration should make clear to Soviet leaders that continued Soviet efforts to press for elimination of tactical-range nuclear armaments could block progress towards reduction both of conventional and tactical-range armaments, with possible negative consequences for the outcome of the START talks and US-Soviet relations generally and, as a consequence, an increase in pressures making for instability in both western and eastern Europe. Since larger Soviet interests are involved than pursuit of the goal of eventually eliminating tactical-range nuclear armaments from Europe, there is reason to believe that the Soviet Union would agree to such a mandate and that it would abide by its agreement.

With more agreement within NATO on long-term aims of the reduction process, and with more clarity within the alliance and with the Soviet Union on the nuclear issue, it may well be possible to move forward to an early first agreement in the CFE talks – and beyond to a second stage.

Appendix

NATO and WTO Air Reductions in the Atlantic-to-Urals (ATTU) Area*

Table 1. NATO Air Forces in the ATTU Area

Location	Aircraft		Number
Belgium	F-16		97* (36)
	Mirage		50
		total	147
Canada (in the FRG)	F-18		36* (18)
Denmark	F-16		52* (26)
	Draken		43* (8)
		total	95
France	Jaguar		127
	Mirage III		106* (26)
	Mirage 5-F		30
	Mirage 2000B/C		58* (45)
		total	321
FRG	AlphaJet		153
	F-4		142* (71)
	Tornado		190
		total	485
Greece	A-7		59
	F-4		50* (15)
	F-5		76* (20)
	F-104		76
		total	261
Italy	F-104		126* (96)
	G-91		141
	Tornado		98
		total	365
Netherlands	F-5		47
	F-16		147* (61)
		total	194

Table 1. (continued)

Location	Aircraft		Number
Norway	F-5		30
	F-16		65* (32)
		total	95
Portugal	A-7		42
	G-91		47
		total	89
Spain	F-5		39
	F-18		46
	Mirage III		48* (24)
		total	133
Turkey	F-4		124
	F-5		80
	F-100		95
	F-104		234* (42)
		total	533
United Kingdom (in UK and FRG)	Buccaneer		52
	Harrier		51
	Jaguar		108
	Tornado		293* (36)
		total	504
United States (in Europe)	A-10		108
	F-4		36
	F-16		328* (96)
	F-111		140
		total	512
	NATO	total	3,814

*

NATO 15% Reduction	707	NATO 50% Reduction	1,987
New Joint NATO/WTO Level	3,107	New Joint NATO/WTO Level	1,827

Table 2. WTO Air Forces in the ATTU Area

Soviet Union

Location	Aircraft			Number
USSR, GDR	Fencer	Su-24		315
Poland	Fishbed	MiG-21		(90)*
(including	Fitter	Su-17,20,22		225
HQ Legnica),	Flogger	MiG-27		135
and Hungary	Frogfoot	Su-25		45
			total	810
Baltics Fishbed	MiG-21			(90)*
Byelorussia	Fitter	Su-17,20,22		45
(including	Flogger	MiG-27		180
HQ Smolensk),	Frogfoot	Su-25		45
and	Backfire	Tu-26		120
Carpathians	Badger	Tu-16		160
	Blinder	Tu-22		120
			total	760
Leningrad MD	Fishbed	MiG-21		(90)*
and	Fitter	Su-17,20,22		45
Odessa MD	Flogger	MiG-27		90
	Frogfoot	Su-25		45
			total	270
Kiev MD	Fencer	Su-24		180
(including	Flogger	MiG-27		45
Vinnitsa)			total	225
Trans-	Fencer	Su-24		45
Caucasus MD	Fitter	Su-17,20,22		225
	Flogger	MiG-27		135
	Frogfoot	Su-25		45
			total	450
		Total Soviet Union		2,515

* Total Soviet Union	2,515
Total Eastern Europe	1,645
Warsaw Pact Total	4,160

Warsaw Pact total after deleting unilateral soviet reduction 3,655

WTO 15	New Joint		New Joint
NATO/WTO	3,107	NATO/WTO	1,827
Level		Level	

Table 2. (continued)

	Eastern European States			
Location	Aircraft			Number
Bulgaria	Fishbed	MiG-21		(110)*
	Flogger	MiG-27		45
	Fresco	MiG-17		15
	Frogfoot	Su-25		45
			total	215
Czechoslovakia	Fishbed	MiG-21		70*
	Flogger	MiG-27	40	
	Frogfoot	Su-25	40	
	Fitter A	Su-7	45	
			total	395
GDR	Fishbed	MiG-21		(225)
	Fitter	Su-17,20,22	35	
	Flogger	MiG-27	25	
			total	285
Hungary	Fishbed	MiG-21	(45)*	
Poland	Fishbed	MiG-21	(360)*	
	Fitter	Su-17,20,22	125	
	Fitter A	Su-7	30	
		LIM 6	70	
			total	585
Romania Fresco	MiG-17		85	
	Orao	IAR-93	35	
			total	120
		Total Eastern Europe		1,645

Notes

We have included in the reduction base all ground attack aircraft, fighter-bombers, and medium bombers, as well as multi-role aircraft which can be used for ground attack. We have excluded air-to-air fighters, training and air defence aircraft. For NATO the total aircraft not counted are: 197 F-4, 39 F-5, 120 F-15, 235 Mirage F-1, 22 Mirage IV and 72 Hawk. For the WTO the following were not counted: 1070 MiG-23, 360 MiG-29, 90 MiG-25 and 45 Su-15.

For the purposes of the reduction purposes set, multi-role aircraft normally classed as fighter aircraft have been added to the category of attack aircraft in this chart, and are indicated by asterisks. The number of aircraft of each type shifted is shown in parenthesis. Stored aircraft are included in the totals.

To arrive at the Warsaw Pact total, we have added to the WTO Fighter-bombers total of 1445 the 1035 multirole MiG-21s from the fighter-interceptors column, the 400 Soviet medium range bombers from the Smolensk Air Army based in European USSR and the 225 aircraft located in the Kiev military district.

According to Gorbachev's reduction proposal 800 Soviet combat aircraft of unspecified type would be unilaterally withdrawn. Western European members of the Warsaw Pact have announced unilateral reductions of 210 aircraft. We have arbitrarily subtracted one-half of the total 1010 aircraft or 505 aircraft from the Warsaw Pact reduction base set forth here.

The NATO proposal as of July was for a 15 per cent cut in combat aircraft. Applying this proposed cut to the reduction base we suggest here the WTO reducing 548 aircraft. NATO must reduce by 707 aircraft to meet the new WTO level. In a 50 per cent reduction, the WTO reducing by 1827 aircraft, NATO must reduce by 1987 aircraft to meet the new WTO level.

Main Sources for this Appendix

The International Institute for Strategic Studies, "The Military Balance", 1988–1989 London: IISS, 1988

Conventional Forces in Europe: The Facts, NATO November 1988; "US Ground Forces and the Conventional Balance in Europe", Congressional Budget Office, June 1988

Problems and Prospects for Negotiations on Short-Range Nuclear Forces

Jane Sharp

1. Introduction

Both the North Atlantic Treaty Organization (NATO) and the Warsaw Treaty Organization (WTO) have been in the process of modernizing their nuclear forces in Europe since the 1950s when American and Soviet forces were first deployed there. Parallel with these changes in the weapons stockpiles have been various proposals for arms control and for nuclear weapon-free zones; usually made by the WTO to pre-empt the latest NATO deployments. The two alliances did not field symmetrical nuclear arsenals since they were dealing with different security needs. This makes arms control very difficult unless, as was the case with the Intermediate Nuclear Forces (INF) Treaty, an entire category of weapons can be eliminated and equal ceilings set at zero. This would be the preferred option of WTO, but NATO leaders consider retention of some nuclear weapons in Europe necessary to compensate for the geographical advantages of WTO. NATO used to claim that its nuclear weapons were to compensate for overwhelming superiority in WTO conventional forces, but since an agreement arising from the negotiations on Conventional Forces in Europe (CFE) could set equal ceilings on key elements of both side's forces, the rationale for nuclear weapons has changed. This point is repeated several times in the Comprehensive Concept Document published at the NATO summit held on 29–30 May 1989.

The INF agreement, singed in December 1987 and ratified in 1988, imposes a global ban on all land-based US, Soviet and FRG missiles with a range between 500 and 5000 km. For NATO this meant the elimination of ground-launched cruise missiles, and Pershing IA and II missiles. For the Soviet Union it meant a ban on all SS-4, SS-5, SS-23 and SS-20 missiles. The Treaty set equal ceilings of zero but involved highly asymmetrical cuts; far more Soviet than American systems had to be destroyed (1752 vs 859).

2. The Impact of the INF Treaty on NATO Modernization

The INF Treaty stemmed directly from FRG concern that the bilateral SALT Agreements did not impose limits on the Soviet IRBMs targeted on Western Europe. In December 1979 NATO endorsed a double-track decision to deploy new INF systems in Western Europe and at the same time to try to negotiate limits on these systems. The new cruise and Pershing II missiles were designed both to compensate for the new SS-20 missiles deployed in the Soviet Union and also to perform the deep interdiction mission of NATO aircraft now vulnerable to sophisticated Soviet air defences.

Parallel with the negotiation of the INF Treaty in the early 1980s, NATO made a number of decisions on modernizing its stockpile of nuclear weapons in Europe. Integral to the December 1979 double-track decision was the unilateral withdrawal of 1000 nuclear warheads from Europe and a one-for one replacement of older warheads by the 572 new warheads on Cruise and Pershing II missiles. In October 1983 NATO's Nuclear Planning Group (NPG) meeting in Montebello, Canada, agreed to withdraw another 1400 warheads and to modernize the remaining arsenal. In October 1986 in Gleneagles, Scotland, the NPG drew up new General Political Guidelines (GPG) for the use of nuclear weapons, to replace the 1969 Provisional Political Guidelines (PPG) and the 1970 General Release Guidelines (GRG). The GPG reaffirmed the policy of Flexible Response, namely that NATO would defend itself against attack in three phases: "direct defence", "deliberate escalation" and "general nuclear response". In anticipation, not of a double-zero INF Treaty, but of the deployment of Cruise and Pershing II, and the retirement of many older shorter-range systems, the GPG placed a higher priority on intermediate-range nuclear weapons that can threaten Soviet targets from Western Europe than on shorter-range systems that threaten Europeans [1].

In effect, these priorities were rendered obsolete by the INF Treaty concluded the following year that eliminated all Soviet and American intermediate-range missiles, retaining only those systems with a range of less than 500 km. The need to do something about limiting shorter-range nuclear forces (SNF) was recognized as early as June 1987, when the NATO Foreign Ministers, meeting in Reykjavik, agreed to seek, in conjunction with a chemical weapons ban and the establishment of a conventional balance, "tangible and verifiable reductions of American and Soviet land-based nuclear missile systems of shorter range, leading to equal ceilings" [2]. During 1988 NATO focused most of its energies on settling the mandate for the new talks on CFE. But Chancellor Helmut Kohl also proposed separate talks to reduce short-range nuclear forces, as West Germans across the political spectrum began to emphasize the singular threat to both Germanies of the short-range nuclear weapons remaining on their territories after the INF agreement. As Volker Ruhe put it: "The shorter the range the deader the Germans". He and other West German parliamentarians were particularly upset by the NATO WINTEX command-post game in March 1989, when the Supreme

Allied Commander Europe (SACEUR) overruled a request by the Chancellor not to fire nuclear weapons at East European targets.

3. Recent Modernization of Soviet Nuclear Forces in Europe

A bone of contention in the current NATO debate on SNF has been the assertion that NATO must modernize, if only to keep up with Soviet modernization. Foreign Minister Eduard Shevardnadze, and other senior Soviet officials, have repeatedly denied that the Soviets are modernizing in this category. They emphasize instead the recent announcement of a cut of 500 warheads in Eastern Europe and the offer to eliminate all nuclear warheads from the WTO portion of the Atlantic to the Urals (ATTU) zone, if NATO will do the same. The 500-warhead cut announced in May 1989 includes 166 air-launched bombs, 50 artillery shells and 204 missile warheads.

In any event, NATO claims these cuts are minuscule compared with NATO cuts since 1979 and cite the following instances of Soviet modernization in the SNF category:

- Old FROG missiles are being replaced by SS-21 (Scarab), specifically, that the number of SS-21 launchers has doubled in the Gorbachev years [3]. This replacement began in 1978 and is presumably drawing to a close.
- Older towed nuclear-capable artillery is being replaced by new self-propelled artillery including 152 mm, 203 mm and 240 mm calibre guns [3].
- Reloads for nuclear-capable systems deployed in Eastern Europe and unconstrained by the INF Treaty are said to have increased between 50 and 100 per cent in recent years so that each launcher now has three reloads [4].

NATO officials have also expressed concern that Soviet SS-24 and SS-25 missiles are currently being tested at intermediate ranges and might be reassigned to targets formerly covered by the SS-20.

4. Putting SNF Talks on the Agenda

WTO initially wanted SNF included in the mandate for the CFE agenda [2]. As early as March 1988, however, a WTO meeting in Sofia called for separate talks on SNF, and on 12 April 1989 the WTO Political Consultative Committee formally invited NATO to participate in SNF talks parallel to CFE, an idea popular with most West European countries who were not themselves nuclear weapon powers.

The USA and UK, by contrast, emphasized the Montebello Agreement of 1983 to modernize SNF. Since 1983 NATO has already modernized its nuclear artillery by replacing the old W-33 warhead on the 8' nuclear artillery projectile with the new W-79 warhead and is developing a new W-82 warhead for the 155 mm shell. Despite many attempts in the late 1980s to clarify the term "modernize", the allies could not agree on what additional steps to take. The United States and UK obviously wanted to modernize by introducing more nuclear-capable aircraft to Britain, by developing a new stand-off air-launched missile, and by extending the range and scope of the Lance missile to the maximum permissible under the INF agreement.

At the meeting of the NPG in Brussels on 19–20 April 1989, General John Galvin, the SACEUR, presented a Nuclear Weapons Requirement Study recommending unilateral cuts of 1500 more nuclear warheads which, together with cuts under the INF treaty, would bring the NATO stockpile down to approximately 2900 warheads. Galvin's cuts however appear to be conditional on a commitment from the FRG to accept a longer-range follow-on to Lance [6].

A follow-on to Lance, which is essentially a new longer-range missile, rather than, for example, updating by shifting from liquid to solid fuel to cut down the reload time, is politically unacceptable to many of the NATO allies, especially West Germany. Despite implying his support for Lance modernization to senior officials in President Bush's administration earlier in the year, domestic opposition made Helmut Kohl want to put off any decision about modernization until after the FRG elections in 1990. The German Social Democratic Party (SDP) wanted a triple zero that would eliminate all short-range land-based nuclear weapons, and Foreign Minister Hans-Dietrich Genscher took a middle position, asking for an agreement on SNF to very low equal ceilings.

On 21 April 1989, immediately after the NPG meeting, a document on SNF, issued by the West German governing Coalition, called for a postponement of any commitment to modernize SNF and for renewed efforts to seek speedy East-West negotiations. This was followed by the visit of Hans-Dietrich Genscher, and the new Defence Minister Gerhard Stoltenberg, to Washington to explain the new West German consensus on SNF [7].

Pressure from Margaret Thatcher in favour of modernization and against negotiation continued. Especially infuriating to the West German government was her version of the "no nukes, no troops" threat made just before the May summit when she asserted that if West Germany refused to accept a new longer-range Lance Great Britain would have to reconsider its commitment of the British Army on the Rhine (BAOR).

The extent of West German irritation with the Anglo-American position was reflected in a speech, made on 26 April 1989 in Copenhagen, by President Richard von Weizsäcker in which he referred to the differences on SNF policy between the continental members of NATO and the others.

President Francois Mitterrand took a middle course, and on a visit to Washington immediately prior to the NATO summit tempered the West German enthusiasm for negotiation and the British enthusiasm for modernization. France is enthusiastic about neither.

NATO looked to Washington for leadership as the bitter Anglo-German dispute over SNF threatened to disrupt the NATO summit scheduled for late May. In the event, President Bush adopted something close to the French position; taking account of Chancellor Kohl's concerns about early modernization and Prime Minister Thatcher's anxiety about early negotiations, he proposed that NATO SNF should be neither modernized, nor negotiated about, until success had been demonstrated at the CFE talks. The NATO allies endorsed this proposal and the communique after the summit specified that SNF talks cannot begin until a CFE has been signed and implementation has begun, and further, that implementation of an SNF agreement cannot begin until implementation of a CFE agreement has been completed.

5. Issues in the SNF Negotiations

Assuming that SNF negotiations are on the agenda, what are the important issues to be decided? When are talks likely to begin? Who should participate? Which systems should be cut? In what geographical zone? And to what levels?

How soon could talks begin? One problem for NATO has been fear that the WTO seeks a denuclearized Europe. Many NATO leaders agree that there are too many nuclear weapons in Europe and would be happy to settle for less, but then fear that once talks began it would be hard to resist pressure for another zero agreement like INF. The Soviet leadership now seems more sensitive to this and on several occasions Mikhail Gorbachev, Eduard Shevardnadze and Victor Karpov have suggested that an SNF agreement could set low ceilings rather than zero. In his speech to the Council of Europe in Strasbourg in early July, for example, Gorbachev suggested immediate talks of experts (that is short of formal negotiations) to discuss optimum levels for a first stage agreement. Such experts talks could begin immediately outside formal SNF negotiations.

The NATO position on when SNF should begin has already been noted. All depends on how soon a CFE agreement can be signed and implementation begun. A simple first-state agreement setting ATTU limits on Tanks, Armoured Personnel Carriers (APCs) and Artillery would best serve the purpose of early SNF talks. Many suspect this is the reason the new NATO proposals are broadening and complicating the agenda, the kind of delaying tactics we used to ascribe to France.

Who should participate in SNF talks? Should SNF talks be bilateral, between the superpowers or should they also involve European states hosting Soviet and US weapons and the other European nuclear weapon powers and France?

The USSR excluded UK and French systems from the INF Treaty on the understanding that these systems would be dealt with later. In the SNF context WTO initially wanted to include France and UK and French and British systems, but they appear to be backing off again. This is reminiscent of the Soviet position on forward-based aircraft at SALT, in which the Soviets several times raised the issue only to drop it in the interest of reaching agreement on other systems. The issue of French and British systems will become more urgent for the Soviet Union as the strategic systems of the superpowers are reduced at START and as the British and French systems are built up. Note that official WTO data presented in January 1989 excludes French and Soviet medium-range bombers, while including all tactical aircraft of UK and France. This may imply that the Soviets intend to ask for cuts on UK and French systems at START.

Which systems should be covered in SNF talks? NATO clearly wants to consider only land-based missiles with a range under 500 km, but depending on how aircraft are dealt with at the CFE talks, the WTO may press for nuclear-capable aircraft to be included. WTO may also press to include naval SNF.

Which geographical zones should be covered in SNF talks? The question is whether talks should deal with the ATTU area, as in the CFE talks, or whether a global approach should be adopted as happened in negotiations leading to the INF Treaty.

To what levels should SNF be reduced? Considerable confusion exists about what constitutes the first, second, third and fourth zero in European nuclear forces. Hugh Beach has clarified the forces involved as follows:

	USA-NATO	Range (km)	USSR-WTO
1st zero	Pershing II	1000–5000	SS-20
2nd zero	Pershing 1A	500–1000	SS-12 + SS-23
3rd zero	Lance	50–500	SCUD
	(Pluton and		
	Hades)		FROG + SS-21
	ATACMS		
4th zero	MLRS	0–50	Artillery
	Artillery		

Beach argues for a third global zero in land-based systems, whatever their munitions, in the 50–500 km range, on the grounds that it would reduce the deep interdiction threat each side poses to the other. He claims the terms of trade are very favourable to NATO. The Soviets would have to eliminate far more, and far more modern missiles, than NATO. NATO would be trading away old systems like Lance and Pluton, and foregoing new systems not yet deployed like nuclear FOTOL and conventional ATACMS. Not having to face the new French Hades and the new US ATACMS and FOTOL would be the primary benefit to WTO of an SNF agreement.

References

1 L. Rühl, *The Nuclear Balance in the Central Region and Strategic Stability in Europe*, NATO'S 16 Nations, August 1987, pp. 18–25
2 NATO communique, 11 June 1987, reprinted in *Survival*, September-October 1987, pp. 461–62
3 Remarks of Richard Cheney, US Secretary of Defense, at the NATO NPG meeting held in Brussels in April 1989, reported by David White, *Financial Times*, 19 April 1989. See also US Information Service, 20 April 1989
4 J. McCausland remarks at meeting of UK Council for Arms Control, 1989
5 See contributions by Jane Sharp to Stockholm International Peace Research Institute. *World Armaments and Disarmament: SIPRI Yearbook, 1988* (Oxford, 1988) and *World Armaments and Disarmament: SIPRI Yearbook, 1989* (Oxford, 1989)
6 See D. Fourquet in *Jane's NATO Report*, May 1989
7 See P. Pringle et al in *the Independent* (London), 25 April 1989
8 Hugh Beach, *The Case for a Third Zero*

The Ideological Dimension of European Security

Laszlo Valki

According to President George Bush, the Cold War started in Eastern Europe and it will come to an end there, too. Indeed, reading again Winston Churchill's Fulton speech of 1946 – which marked the beginning of the Cold War – we have to acknowledge that the British politician spoke almost exclusively about the fate of Eastern Europe. The speech was a clear message to Joseph Stalin and also to Western political leaders. Nevertheless, it did not change the direction of developments at all. Responding to Churchill, Stalin declared that the Soviet Union had a legitimate right to be surrounded by "friendly" states on the continent. Although this right has never been officially recognized by the West, neither has it really been challenged. The Truman doctrine and the Containment policy did not affect Soviet intentions towards the East European countries.

I am not going to discuss here the definition of Cold War, or the question how long it did or will last. The appropriate question is whether the recent dramatic changes in Eastern Europe will bring about an end to the historical hostilities between East and West? The answer is probably no. The origins of hostilities can be traced back to the turn of the century when the Communist movement claimed it had found a general panacea for the diseases of society: exploitation, unemployment, crime, economic stagnation and depression, nationalism, colonialism and war. The strong belief of the late Communist leaders and ideologists that they could heal the troubled society is understandable. The diseases were in fact fatal and their ideology had a clear logic and a serious scientific basis provided by great European thinkers.

The First World War justified the Communist diagnosis, as later did the Great Depression, the birth of Fascism and the Second World War. There were, however, two major contradictions in the Communist ideology. First, the panacea – the new type of socio-political system – was not suitable for healing society and the original ideas were tragically distorted by Stalinism. Second, the members of the Communist movement always thought and – in accordance with the real possibilities – acted in terms of a worldwide revolutionary process. In other words, the ideology was inherently expansionist, without having the necessary social, political, economic and military strength behind it.

Now we can observe in the East a tacit or partial, and in some cases an explicit and full recognition of these contradictions. This recognition was translated into foreign policy actions by the new Soviet leadership. Since hostilities had shifted in the meantime to other continents, President Mikhail Gorbachev knew that he also had to act outside Europe. The withdrawal of foreign troops from Afghanistan, Angola and Cambodia are the most important steps taken or to be taken in this direction. They are being accompanied by active measures in Europe, such as the unilateral reduction of armed forces, and by "passive" measures such as the renunciation of the "Brezhnev doctrine". Thus it may well be that while the Cold War started in Eastern Europe the end of hostilities can be achieved only worldwide, and especially on the sixth continent, namely the Continent of Ideology.

Historically, the changes have already reached the point of no return. Practically, however, some might come to a different conclusion. In any case, tremendous wisdom is needed on the part of all actors in international relations, regardless of how they see the shape of the world in the coming century.

Non-Military Aspects of European Security

Klaus Gottstein

Modern technology has made the world small and interconnected. Europe is even smaller, and it has very close links, in many respects, with all parts of the world. If one wants to travel from one African capital to another African capital, the fastest way is usually via London, Paris or Frankfurt. European security can only be seen within the global context. Europe cannot remain unaffected by catastrophes in other parts of the world. On the other hand, major catastrophes in Europe would also have dire consequences all over the globe.

It is not difficult to imagine the occurrence of such catastrophes. The Third World is a powder keg because of so many unresolved problems of poverty, hunger, overpopulation, inflation, indebtedness, dictatorship. Europe is a powder keg because of the presence there of thousands of nuclear weapons, and because it is divided by a frontier between two different social systems which are in a state of uncertain transformation and reform.

Ever since Hiroshima and Nagasaki the task of avoiding another application of these horrible weapons has become of paramount importance. Since nuclear weapons cannot be disinvented this task will remain a permanent one. The Pugwash Movement has given this task top priority from the very beginning, and with apparent success: nuclear war has been avoided so far. Nuclear weapons, however, proliferated. Pugwash had no success in its attempts to ban the production, testing and deployment of nuclear weapons. The reason for this failure was, and still is, the mistrust of governments with respect to the potential long-range political and military goals of the other side. Deterrence was, and still is, considered irreplaceable. Even though there may be detente, cooperation, and even reasonably friendly relations between East and West, as at present, no leading politician on either side is prepared to guarantee that there will not be a new threat ten years from now which would require renewed deterrence. Considering the long lead-times for the introduction of new weapon systems, there is great reluctance to destroy the majority of existing weapons. The only way to change this situation would be to create trust, on both sides, with regard to the long-range intentions of the other side.

In the past this was impossible. In a situation in which one side considered "peaceful coexistence" as the best way to continue class struggle for the victory of communism, and the other side saw "peaceful coexistence" as the only available way for "tearing out the evil page of communism from the book of history", there was no chance for creating trust by discussing long-range goals. There was only a chance for discussing common interests if long-range goals were not mentioned. This meant that only short-range problems of crisis prevention and crisis management, in which both sides were interested, could be discussed, notwithstanding their deep-seated mistrust of each other. Significant progress in arms control and disarmament was not feasible under these conditions.

This situation has changed under *perestroika* and *glasnost*. Old enemy images are beginning to fade away. The prospect of real progress, based on mutual acceptance of the long-range goals of the other side, starts to appear above the horizon. But there is still a long way to go. A relapse into the old enemy images cannot yet be excluded. In the preceding paper by Laszlo Valki, it is suggested that the recent dramatic changes in Eastern Europe will probably not bring about an end to historical hostilities between East and West. In my opinion this is precisely an "essentialist" perception. Essentialists believe in the necessity of continuing confrontation between East and West. Quite consistently, it is explained in the paper that the belief of the Communist movement (that it had found a general panacea for the diseases of society) had a clear logic and a serious scientific background, and was justified by historical events, although it does not mention that already at the beginning of this century, Socialist thinkers, like Bernstein, had abandoned the idea of revolution and had come to believe in a transformation of society by reforms. Valki states that members of the Communist movement have always been thinking and acting in terms of a worldwide revolutionary process. He notices, however, two contradictions:

- the panacea was not suitable for healing society, having been tragically distorted by Stalinism; and
- the ideology was inherently expansionist without having the necessary social, political, economic and military strength behind it.

It will be necessary to overcome such expectations of continuing hostilities. Scientists and scholars can have an important part to play in this task because of their international connections, their influence, their knowledge, their methods, and their level-headedness. Whereas during the Cold War the task was the prevention of nuclear war, today's task is the creation of trust by projecting common goals, and the removal of mistrust by studying the perceptions of long-range goals of the other side and by the other side.

As an example of how this can be done, let me mention briefly a conference [1] recently held near Munich with the participation of nine Soviet experts on the West and ten American Sovietologists. Representatives of different schools of thought were invited intentionally. It turned out that both in the Soviet Union and in the United States it is possible to identify three groups of perceptions. Their holders have been classified as essentialists, mechanists and interactionists. The essentialists consider the other side as a principally hostile power which can only be met by confrontation. The mechanists see it as a geopolitical opponent governed by opportunism with which it is best to negotiate from a position of strength. The interactionists believe in peaceful competition and constructive cooperation.

At this conference it was impressive to listen to the arguments of Soviet and US interactionists. But it also became clear that essentialists and mechanists on both sides still exist, waiting for the failure of present attempts for reform and East-West cooperation. The view of Western essentialists and mechanists is that the present expression of willingness by the Soviet Union to cooperate with the West is dictated by President Gorbachev's need to gain time for his economic and social reforms. Once the Soviet Union has gained strength through those reforms Western sceptics fear Soviet ideology will return to its natural expansionism. Therefore, the argument goes, it would be unwise for the West to disarm prematurely. It might even be unwise to help the Soviet Union in its efforts to get out of its economic impasse and gain strength.

In Pugwash, of course, we are neither essentialists nor mechanists but interactionists. We believe in the value of cooperation as such for the preservation of peace. But we should take into account the fact that the essentialists and mechanists on both sides still exist, waiting for their chance to be proven right. This means that we should study their arguments and look for appropriate ways to deal with them. The future of disarmament can only be seen as reasonably secure if the arguments leading to the mistrust of the essentialists and the mechanists can be fully understood and dismantled at their foundations. This will require convincing evidence *shown by the other side*. It is not clear to what extent and how soon this will be possible but this is where the next tasks of our generation lie.

As I said on another occasion [2], I think Pugwash should enter this field. As in all the earlier years, Pugwash deals mainly with the symptoms and consequences of East-West confrontation, not with the possibilities for overcoming the basic motivations for this confrontation. Now it has become feasible to enter the field of studying, and changing, these motivations. The chance should not be lost.

The only hope for avoiding catastrophe lies in finding a controlled path – under the provisions of the scientific-technological age – from the present to the future of mankind. This does not mean that the products of science and technology in the form in which they were developed in Europe, the United States or Japan should be imposed on every village in the world. On the contrary, great care should be taken to preserve as much as possible the cultural identity of individual regions. Europe has a special responsibility here because it is its cultural identity that is unconsciously reflected in many products of Western civilization, and exported, often in a thoughtless way.

Mastering the future of our overpopulated planet with its problems of pollution of air, soil and water, deforestation, spreading of deserts, shortage of resources, abundance of arms, enormous social problems deriving from the vast gaps between the living standards of the rich and the poor within countries, and between rich and poor nations, will require a new ethics of mutual assistance and burden sharing. Those who voluntarily forgo the chances of improving their lot by cutting the rain forest will have to be compensated by those who already live in affluence in an industrial environment. Those who encouraged the poor countries to accept loans for industrialization projects which turned out to be unprofitable or impracticable due to poor planning or lack of foresight, should now accept their share of the responsibility and of the financial loss, by waiving part or all of the corresponding debts.

The tasks facing mankind are so great and so complex [3] that not only the political but also the scientific, cultural, economic and social cooperation of all nations is required. The poor, in their daily struggle for survival, cannot be expected to come up with the sophisticated solutions required for taking into account the complexities of the world system. It is the rich who have to shoulder this task, because they have the required resources at their disposal. They have the affluence giving them the leisure and the facilities needed for work of this kind.

Cooperation in joint projects is not only necessary in a technical sense for the solution of the problem mentioned, it is also needed from a psychological point of view: it creates a feeling of togetherness, thereby overcoming enemy images. Such enemy images dividing East and West and North and South have, fortunately, become less pronounced between East and West in recent years. But they still exist in many quarters and should not be underestimated. They will only be put to rest once a general feeling of mutual trust will firmly have taken hold among the political elites. This trust will mean that worstcase scenarios will no longer be entertained of a type assuming that some opponents in other parts of the world do, or might in the long range, harbour plans for world domination. Joint efforts for common causes are the best way to create mutual trust.

References

1 International Conference on the Mutual Perceptions of Long-Range Goals in the East-West Conflict, Tutzing, May 29 – June 1, 1989. (Proceedings to be published)

2 K. Gottstein, *New tasks for Pugwash, or new Pugwashes for new tasks*? In *Proceedings of Pugwash: Shaping its Future*, A seminar in honour of Joseph Rotblat's 80th birthday, London, November 1988

3 See, e.g. K. Gottstein, *Scientific and Cultural Cooperation as a Means for Reducing East-West Tensions*. In: Joseph Rotblat and Laszlo Valki (eds.), *Coexistence, Cooperation and Common Security*, (MacMillan Press, London, 1988)

Part III

Chemical Weapons and Their Control

Chemical Weapons – Destruction, Conversion or Endless Debate?

Karlheinz Lohs

The Pugwash Movement can claim to have undertaken great efforts to bring about a comprehensive, worldwide ban on the development, storage, proliferation and use of chemical weapons. There is no need to list here the wide range of activities so far undertaken by Pugwash on "Chemical Weapons" and the rich variety of topics discussed in this context at conferences, symposia and specialized workshops. There is hardly any subject of relevance to the ban on chemical weapons to which profound treatment has not been given by Pugwash.

All conclusions deriving from Pugwash activities in this connection have been submitted without delay to the negotiators in Geneva as well as to all the governments concerned and to the UN Special Agencies. It is my belief that Pugwash documentation has had a positive impact on some bargaining positions of the governments involved. Certainly the Pugwash Movement should at least be given credit for having persistently confronted governments and the general public with the urgency of a ban on chemical weapons and for having drawn their attention to the consequences of chemical warfare.

It is, nevertheless, a fact that until now the outcome of all these Pugwash exercises has remained unsatisfactory. Despite all the progress so far made by negotiators in Geneva, we have not yet seen a signed treaty providing in comprehensive terms for a worldwide ban on chemical weapons. We are still witnessing mutual recriminations concerning the absence of a formal treaty. And the use of chemical weapons has continued in regional conflicts, even against the user's own civilian population. Research and development of new chemical warfare agents continues. And the United States persists in producing binary combat substances. Suspicions remain, as do the resulting unrealistic verification demands.

At Pugwash conferences we are still debating technical details relating to destruction, conversion and verification, although we are fully aware that these debates will have to be continued for more decades unless political determination results in a long-overdue resolution banning chemical weapons. For this category of weapons has long been rejected even by a majority of the military, since they are means of mass destruction, are actually aimed primarily at civilian populations and are, consequently, without real strategic importance for warfighting.

Continued meticulous expert discussion of details, therefore, has become almost meaningless. It is no longer essential to be alert about possible new methods of verification. For the existing knowledge and available equipment are sufficient. The same applies to ways and means for detoxification and for the destruction of chemical warfare agents, as well as for conversion of initial and intermediate products. Nor can we be satisfied by the growing complexity of draft conventions.

An agreement among all UN member states on the following terse formula would be quite sufficient:

> Any military use of poisons against human beings and the environment is absolutely banned under all circumstances. Offenders against that ban would isolate themselves from the community of nations and shall be deprived of all social, economic and other inter-state contact.

Such a declaration will be easily understood by anyone and is, therefore, conveniently verifiable.

Pugwash should undertake a determined commitment to make the community of nations understand that even the most sophisticated problems can be resolved in a simple, straightforward manner, provided the political will exists.

Destruction of Chemical Weapons

Sven Lundin

The preceding paper by K. H. Lohs contains an interesting suggestion for a programmatic activity, namely that Pugwash should induce the UN to accept unanimously a declaration about abstaining from any use of chemical weapons, with the understanding that once all nations had accepted such a resolution and suggested sanctions, no further measures would be needed.

The suggestion referred to is a rather principal approach that relates not only to the question of the destruction of chemical weapons under a future Chemical Weapons Convention (CWC). The question arises, does the concept imply that a CWC would then not be necessary? Further, what status would the 1925 Geneva Protocol have in this context? And, in what way would it be easier to muster the necessary political will by this approach than for a CWC?

I do not agree, on a principal level, that the longtime efforts of Pugwash regarding Chemical Weapons have been made in vain, just because, so far, no Chemical Weapons Convention has materialized. On the contrary, I think that these efforts have been necessary – and successful – in picking up all the real or alleged difficulties to obtain a CWC, which have been put forward by those genuinely concerned about the viability of, or actively opposing, the CWC. This has made it possible to continue a discussion about ways to overcome such dilemmas. These issues have shifted over the years, due to technical developments, but also and not least, due to fundamental changes in the political attitudes. It seems to me that the constant preparedness by Pugwash to interact actively in this discussion has been invaluable and should not be dismissed under any circumstances, even if an all-out attempt were undertaken to get the problem solved once and for ever. Thus, the deliberately pessimistic views have not been allowed to stand unchallenged and result in a cessation of the efforts to get a CWC. This is a laborious and time consuming task but it has to be done and with all the competence that can be mustered.

Would it be futile to discuss such matters as the verification of the destruction and conversion of chemical weapons, because existing knowledge is sufficient? I do not think so.

In whatever form a global agreement to get rid of chemical weapons might in the end appear, no nation concerned with its security and subject to threats of use of chemical weapons either from the superpowers or from regional powers, would be content to accept a mere declaration regarding performed destruction of the threatening chemical weapons. These nations would like to have guarantees that the destruction was carried out. An easy way of doing that is to allow the concerned parties, or, for them, the inspectors of the Technical Secretariat, to watch the destruction. Such a measure might have been considered impossible a few years ago but this is not the case today.

In fact, verifying the destruction of declared stockpiles today might offer the least difficult technical problems under a future CWC. On this there seems to be a general understanding in the Conference on Disarmament (CD). On the other hand, the detailed technical procedures, even if outlined in principle in the "rolling text" of the CWC, have not been discussed sufficiently to give all concerned Parties an acceptable understanding of how these methods might work and what security they might provide. If Pugwash experts could contribute in this field with concrete and applicable measures, this would be most helpful, not least in the context of the work of a future Preparatory Commission. I do agree, though, that it might not be necessary to burden the text of the CWC with further details.

There remain, however, two problems, which are of a political rather than a technical nature.

One concerns the "order of destruction", i.e., in which order should stockpiles be destroyed in order to keep undiminished security *vis-á-vis* potential adversaries. This question may in the first place be a problem for those nations which have large stockpiles of chemical weapons and their, perhaps neighbouring, potential adversaries. It will probably be necessary to let these nations suggest tentative solutions, which can then be analysed and hopefully agreed upon by the negotiating countries and other interested nations. Efforts in this direction have been reported in the CD.

Further, one should not overlook the possibility that a substantial destruction may be performed during the period between the signing and entering into force of the CWC, when the weapons to be destroyed under CWC will have to be declared. I am not sure to what extent Pugwash efforts really can contribute in an area, in which national security concerns may obscure, or make it impossible, to obtain, in advance of a CWC, sufficient knowledge as a basis for a meaningful discussion.

The second, and principally more genuine uncertainty, relates to the difficulty, not to say impossibility, to verify technically that no hidden stockpiles exist in a country Party to the CWC, that is, stockpiles which were deliberately not declared and destroyed as the CWC would demand.

First, in case no such stockpiles are actually hidden, it would of course be impossible to verify that fact, unless every potential site had been scrutinized, which would be practically impossible.

Second, if there were actually hidden stockpiles, the same practical difficulty to scrutinize all potential sites would exist. Further, no technical methods to find out about them would exist, unless there were other indicators resulting from outright espionage or defectors, or perhaps occurring environmental damages due to the toxic content of the stockpiles. Only then might perhaps national technical means of verification be useful in order to underpin complaints to the future Technical Secretariat. From a political point of view, it would be necessary to decide to what extent a nation could live with this uncertainty under a CWC. An analysis of this problem would take us far into actual scenarios and political conditions in tentative Parties to the CWC and is beyond my scope.

The fact that hidden stockpiles cannot be disclosed with any degree of probability must be put in relation to what sort of threat such hidden stockpiles might actually pose to hesitating subscribers to the CWC. It would be interesting to discuss what sorts of threats limited amounts of hidden stockpiles of chemical weapons might constitute over a period of years. Further, over time, what risks and costs would a State Party suffer, when his hidden stockpiles were getting obsolete and deteriorating. Parts of the answer to this question have surfaced over the last decade, both in the USA and the USSR with respect to the problems of handling old stockpiles of chemical weapons. Also, what implications might appear for the CWC if hidden stockpiles of chemical weapons were found to exist in a country Party to the CWC? One should then again consider developments over time, not only the hypothetical moment when a sort of Chemical Warfare capability suddenly arose, because a limited amount of usable chemical weapons were instantly available.

I should make clear that the problems referred to above relate to a varying extent also to production facilities and to conversion of chemical warfare agents into degradation products which cannot, again, be immediately used to produce the same chemical weapons.

In conclusion, I think it is to the merits of Pugwash that the Movement early realized the need to address not only the overshadowing question of nuclear weapons but that it also has put considerable effort and thoughts into solving the problems concerning the other weapons of mass destruction, chemical and biological weapons. In fact, it is with respect to these systems of weapons of mass destruction that disarmament efforts have been most successful so far. Pugwash should not disregard this achievement to which it has contributed profoundly.

An International Scientific Methology for Verification of Non-Production of Chemical Weapons

Nikolai Enikolopov and Kirill Babievsky

1. Introduction

The majority of nations warmly agree that the cessation of the operation of every chemical weapon production facility, including those of private enterprises and transnational corporations, should be ensured by means of strict verification. This should include: systematic on-site inspections; verification of the accuracy of declarations; sealing by inspectors of the facility to be closed; periodic checking of the preservation of seals up to the moment when the seals are removed; the destruction or the examination of such facilities by the representatives of the national verification organization; and the preparation of reports from the international inspectorate for transmission to the Consultative Committee.

For the purpose of effective verification of the destruction and dismantling of chemical weapon production facilities, the Soviet Union proposes that provision should be made for systematic international on-site inspections and that a procedure should be worked out for visits by international inspectors, who would be present at chemical weapon production facilities. Final international verification would be carried out upon the full termination of the process of the elimination or dismantling of the entire facility. Conversion of facilities for the destruction of stocks of chemical weapons, as well as their elimination upon the completion of the destruction of stocks, should also be carried out under supervision of the international verification personnel in closest cooperation with the national verification system.

A wide range of sensors and equipment is currently available to the chemical industry, necessary to monitor efficiently process parameters at plants. At the same time, the development of monitoring methods should include new, more sensitive and selective sensors. We are here primarily concerned with microsensors, which can detect reagents both in gaseous and liquid environments. Small, simple, cheap and reliable chemosensors combined with microprocessor technology should produce the best results for verification systems.

In recent years a number of chemical substance sensors for effective industrial installation control have been developed in different scientific laboratories. The so-called SAW-chemosensor is one of them. It is known that as far back as 1964, W. H. King [1] was the first to suggest the use of acoustic waves in piezoelectric sorption detectors. The same detection technique was later applied by H. Vohltjen and R. Dessy [2] to surface acoustic waves. Under this principle the chemosensor's gas-selective surface is used to convert the chemical signal into a corresponding electric pulse (chemical interface). The final features (selectivity, reversibility, sensitivity and stability) depend on the type of reaction and the morphology of the gas-selective surface. The existing model was initially developed in the Netherlands [3] for NO_2 analysis. The system can function both with selective membranes and a biochemical interface (the ACAL system) using ferment inhibition. Systems with chemoresistors and adsistors, chemosensitive field-effect transistors [4], as well as surface potential and other physical measurement systems, can also be used as microsensors [5].

New sensory systems are also widely available that form the basis for high resolution gas and liquid chromatography detectors: electro-chemical cells, enzyme detectors [6], photo-ionization detectors [7], mass-spectrometric detectors of various kinds [8], including those for the mobile MM-1 mass-spectrometer [9]; receptor- [10] and antibody-based biosensors [11].

2. Analytical Instruments

State of the art analytical instruments are at the heart of both mobile and stationary (central) laboratories designed to control and monitor chemical weapons non-production at commercial chemical plants. Tables 1 to 5 show the features of the analytical equipment used in research laboratories in a number of countries to detect hazardous substances of different classes.

Verification analytical instruments have lately been further developed and broadly utilized in various analytical techniques.

Table 1. Mass Spectrometers

Instruments used in developing the analytical procedure	Basic features of used instruments and the coupling system	Analysis of monitored CW compounds
The mobile mass spectrometer MM-1 Bruker-Franzen Analytik GMBH. Kattenturmer Heerstrasse 122, D-2800 Bremen 61, FRG	Glass quadrupole, shock proof. Mass range 1-400 amu. Stability 0.05 amu/day. Sensitivity typically 5 ng in solution or as dust. Mountable in vehicle [9,13] Can be used in SNAL system [28]	See references [27] and [28]
High resolution mass spectrometer FINNIGAN MAT 8200 Finnigan MAT GMBH, D-2800, Bremen 14, FRG	Mass range 1-2100 amu. Direct coupling to a Varian 3700 gas chromatograph and INCOS data system	Organophosphorus CW agents and their precursors, yperite, mycotoxins etc [13,29]
Triple stage quadrupole (TSQ) mass spectrometer FINNIGAN MAT 45 A. Finnigan MAT GMBH, D-2800, Bremen 14, FRG	Used in tandem mass spectrometry with INCOS data system interfaced with a FINNIGAN MAT 961 gas chromatograph	Organophosphorus CW agents, yperite, BZ, CS, mycotoxins etc [13, 29]
High sensitive mass spectrometer JEOL D300. JEOL LTD, 1418 Nakagami Akishima, Tokyo 196, Japan	Mass range 1-4500 amu. Direct coupling with a HEWLETT-PACKARD 5992 B gas chromatograph	Yperite, tabun [30]
Mass spectrometer JEOL JMS-01-SG-2 JEOL LTD, 1418 Nakagami Akishima, Tokyo 196, Japan	With JEC 980 B data system interfaced by direct coupling with CARLO-ERBA 2100 gas chromatograph	CW simulated compounds [31]

Table 2. HRGC Instruments

Instruments used in developing the analytical procedure	Basic features of used instruments and the coupling system	Analysis of monitored CW compounds
Compact HRGC instrument MICROMAT HRGC 412 NORDION Instruments OY Ltd, PO Box 63 SF-02201 ESPOO Finland	Compact capillary gas chromatography with AT, EC, TI and other detectors. The powerful microprocessor controls timing, temperatures and analytical sequences	Organophosphorus CW agents, yperite, BZ, CS [32,33]
High resolution gas chromatograph VARIAN 3700 GC Varian Ass., Instruments Group, 611 Hansen Way, Palo Alto, California, USA	High performance two column HRGC instruments. TC,FI,EC and FP detectors	Sarin, Soman Tabun, VX [34]
High resolution gas chromatograph HEWLETT-PACKARD 5992B Hewlett-Packard Europe, PO Box CH-1217, Geneva Switzerland	Modular HRGC instrument. Capillary columns	Yperite, Tabun [30]
High resolution gas chromatograph FINNIGAN MAT 961 Finnigan MAT GMBH Postfach 144062 D-2800, Bremen 14, FRG	Used in tandem MS/GC system interfaced with FINNIGAN MAT 45A mass spectrometer	Organophosphorus CW agents, CS, Yperite, BZ [13,35]
HRGC instrument BECKER, TYPE 4092 Becker, Netherlands	Capillary gas chromatograph with IT-detector. Obsolete instrument	Organophosphorus compounds with $P\text{-}CH_3$ bond [36-38]
High resolution gas chromatograph CARLO ERBA 2100	Interfaced by direct coupling with JEOL JMS-01 SG-2 mass spectrometer. Capillary column	See reference [31]

Table 3. HRLC Instruments

Instruments used in developing the analytical procedure	Basic features of used instruments and the coupling system	Analysis of monitored CW compounds
Gradient HR liquid chromatography instrument DU PONT 850 HPLC DuPont Scientific Process Instruments Division. Concord Plaza, Wilmington DE 19898, USA	High sensitivity HRLC system with DuPont variable wavelength UV detector, 3-head pump	CS,CN,BZ,CR and CW degradation products Mycotoxins [17,39]
High resolution liquid chromatography instrument KONTRON 640 HPLC. Kontron Co	High sensitivity HRLC system with Kontron 200 processor, Kontron-Uvicon 722 UV-detector and an enzymatic detector (ED). Direct coupling to the automatic analytic system AKEA	VX, Tabun, Sarin, Soman [6,18] see also [40]
High resolution liquid chromatography instrument HP 1090A LC Hewlett-Packard Europe, PO Box CH-1217, Geneva, Switzerland	High sensitivity HRLC system with Rheodyne 7010/7012 Injector and Diode array detector. Coupled with HR 85B Integrator and recorder LiChrosorb Hibar RP-18 columns	Organophosphorus CW agents, yperite, BZ, CS,CR etc. [19,20]

2.1 Mass-Spectrometric Methods

Mass-spectrometric (MS) methods are normally used to analyse a composite compound mix with small concentration of compounds. Electronic- and chemical-ionization-based low resolution mass-spectrometry yields quite satisfactory results when used to screen samples with concentrations of the order of 1–10 grammes per cubic centimetre. Higher resolution can be achieved with high-resolution mass-spectrometers, or tandem mass-spectrometry, in particular with ionization techniques. In compliance with the terms of the proposed convention, an in-depth analysis of the chemical substances to be banned and controlled and the sequential order of mass-spectrometric analysis methods, are now available [12].

To study certain toxic substances in atmospheric samples A. Hesso and R. Kostiainen [13] compared the sensitivity and the resolution of two mass-spectrometric methods of contaminant analysis:

Table 4. NMR Spectrometers

Instruments used in developing the analytical procedure	Basic features of used instruments and the coupling system	Analysis of monitored CW compounds
High resolution Fourier Transform NMR spectrometer BRUKER AM-200PFT. Bruker Analytiche Messtechnik GMBH, Silberstreifen D-7512, Rheinstetten 4, FRG	The microprocessor supported control system, super-conducting magnet, 1H-Frequency - 200 MHz	Trichothecene mycotoxins [21,41]
High resolution NMR spectrometer JEOL GX400 Jeol Ltd, 1418 Nakagami Akishima, Tokyo 196, Japan	1H-Frequency - 400 MHz. Pulse FT NMR spectrometer with PDP 11/23 data system, deuterium lock and dual probe-head	Organophosphorus CW agents, yperite, CS etc. [42]
Fourier Transform NMR spectrometer JEOL FX-90 JEOL Ltd, etc	High performance 90 MHz magnet system	Yperite, tabun [30]

Table 5. FT Infra-red Spectrometer

Fourier Transform Infrared spectrophotometer NICOLET 60 SX FTIR	High resolution FTIR spectrometer. Direct coupling with a HEWLRETT-PACKARD gas chromatograph equipped with a FI detector and a 25m x 0.32mm SE-54 column	Sarin, soman, tabun, VX. yperite [43]

Selected ions monitoring with a combination of gas chromatographic and MS methods. A high resolution bifocal MS instrument was used to this end, and two fixed ions were chosen for CW measurement. Metastable ion monitoring was selected as a supplementary technique for the bifocal instruments.

Tandem mass-spectrometry combined with gas and liquid chromatography. A three-stage quadruple MS was applied here. Both positive and negative ions were analysed. The gaseous phase comprised methane, isobutane and ammonia (for subsidiary ions). The authors demonstrated that both methods are highly efficient in the analysis of a wide range of toxic substances such as organophosphorous CW, yperite, mycotoxins, BZ and CS. The ammonia gaseous phase proved the most sensitive to sarin, soman and tabun adducts. This also facilitates the analysis of organophosphorous compounds.

The experiment determined the requirements for high reliability MS study of the monitored chemicals:

- the instrument used should be able to locate daughter ions and possess a minimum resolution of 5000;
- there must be a chemical ionization feature, and both positive and negative ion detection;
- a direct sample introduction, or a HRLC feature, should also be available;
- the data read-off system should allow for scanning changes in the course of analysis.

Table 4 gives other data on the use of some MS instruments for analysis.

A mobile laboratory for the monitoring of toxic substances and their predecessors "on site" is based on a Bruker-Franzen [14] MM-1 instrument with a special dividing head. The system consists of a gas chromatograph with a sensitive MS detector. It is mounted on a standard minibus chassis and has an independent power supply.

Toxic compounds are identified through library-based EI mass-spectra computer analysis.

2.2 High Resolution Gas Chromatography

Gas chromatography is used to split composite compounds of volatile substances into separate components, which are then analysed with other techniques (e.g. photo-technical infra-red spectroscopy). Two-channel high resolution gas chromatography is a standard technique for preliminary chemical agent identification. Two cores with different stationary phases and detectors permit four separate analyses to be carried out. The following detectors are the most commonly used:

- flame-ionization detector (FID)
- photo-ionization detector (PID)
- alkaline thermo-ionization detector (ATD)
- electron capture detector (ECD)

The so-called RS-spectrometry (retention spectrometry) has been developed as a fast and reliable method for identification and detection of multicomponent CWA systems. The above technique which is an extension of classic chromatography is based [15] on putting the sample through several capillary cores covered with different stationary phases. Here every core gets an equal share of substance (there are six of them normally). At the outlet the divided compounds (in vaporous state) pass through a single common detector. The resultant spectrum of chromatographic peaks is characteristic for each individual compound, because it reflects the nature of the substance's reaction to the various phases. High sensitivity is one of the main advantages of the RS method. Its shortcoming is that it does not offer a sufficient number of different phases.

A Finnish compact and relatively cheap capillary core "Micromat 412" can be used to enhance the operational efficiency of a mobile laboratory that verifies the compliance with the Convention on chemical disarmament.

There is the possibility that an automatic "on-site" monitoring system could be developed to identify CWA on the basis of a three-channel gas chromatograph, equipped with EC, AT and IT (ion trap) detectors [16]. The method was used to describe an effective simultaneous identification of sarin, soman, tabun and yperite from one sample. Nerve gases are AT-detectable in quantities of 1–5 picogrammes (pg), while yperite is EC-detectable at the level of 50 pg. High sensitivity of the order of 100 pg was achieved with IT-detectors. Finnish researchers point to the possibility of using this technique in mobile control laboratories. Table 2 contains published data on the application of gas chromatography instruments to CWA analysis.

2.3 High Resolution Liquid Chromatography (HRLC)

Inverse phase HRLC is currently an efficient means of dividing organic composite compounds, containing a wide range of functional groups. As the technique is particularly suitable for analysing thermally unstable and low-volatile compounds it is successfully applied to the known toxins, their precursors and their hydrolitic derivatives [17].

Recently a highly selective and sensitive method based on the liquid-chromatographic separation of components combined with enzyme and ultraviolet spectroscopic detection has been developed to analyse nerve gases [6], [18]. The technique was successfully applied to quantitative sarin, soman and tabun analysis in atmospheric samples.

DAD-detectors (diode array) [19] open up further possibilities for enhancing the efficiency of the liquid chromatography technique. In contrast to the normal fixed-type detectors and changeable wavelength detectors, DAD allows one to obtain both spectral and multiwave information in a single chromatography. A special gradient elution programme has been worked out to separate and detect a considerable number of chemical weapons and their precursors and degradation products, including organophosphorous nerve gases [20].

At present, many countries produce quite advanced instruments for liquid chromatography. The Soviet Union manufactures a very compact liquid chromatography "Millichrom" device.

2.4 Nuclear Magnetic Resonance Spectroscopy (NMR)

Nuclear Magnetic Resonance (NMR) spectroscopy is a versatile method of structural analysis of compounds soluble to a fairly high concentration. In verification laboratories NMR-spectrometers can be used to identify known compounds; and to obtain an exhaustive structural analysis of unknown compounds either in composites or in pure form.

NMR-spectroscopy seems to be of the most use in monitoring toxic compounds and precursors whose molecules contain phosphorous and fluorine atoms, as well as complex organic compounds. Currently NMR-spectrometers with superconductive magnets routinely use resonance frequencies of several hundred megahertz. This in turn leads to a sharp increase in the sensitivity of the method.

Proton-based high resolution NMR-spectroscopy (the whole of the operation takes about 10 minutes) and carbon-13-based spectroscopy (10–15 hours) permits one to analyse and to describe quantitatively a sample containing, say, 1 mg of trichotecenic mycotoxin in the mould that often covers spoiled products [21]. Recently-developed special methods, such as the 2D NMR technique and highly computerized NMR-spectrometers, significantly broadened the scope of high resolution NMR-spectroscopy, making it one of the most important instruments in a verification laboratory for chemical weaponry non-production at civil chemical plants. Table 4 contains examples of the application of high resolution NMR-spectroscopy to the analysis of toxic compounds. Examples are also offered in some publications [22], [23].

2.5 Infra-Red Spectrometry

Like NMR, infra-red spectrometry is an efficient non-destructive method of obtaining structural information for compounds arrived at through a chromatographic separation of samples. Although Fourier-type infra-red spectrometers (FTIR) are less sensitive than mass-spectrometers, they permit a reliable structural identification of chemical samples at the pg level. However, a considerably higher sensitivity is achieved if Fourier infra-red gas chromatography is combined with the matrix technique. The chromatographic elutriator is frozen into an argon matrix at 12K and is then analysed on a Fourier infra-red spectrometer. The results of research in that direction are quoted in a number of publications [24], [25]. Those publications that deal with environmental pollutants, such as dioxine [26], on the basis of the matrix Fourier infra-red spectroscopy method, are of particular interest.

The techniques and analytical instruments given in the Tables are widely known and used for scientific study and monitoring in ecological laboratories.

References

1 W. H. King jr, *Anal. Chem.*, 36, 1735 (1964)
2 H. Wohltjen, R. Dessy, *Anal. Chem.*, 51, 1458, 1465 (1979)
3 M. S. Nieuwenhuizen et al, *Electronics Letters*, 22, 184 (1986)
4 T. Karpetsky, *Hardware aspects of monitoring compliance*, UN Meeting of experts on verification, Dagomys, USSR, 1988

5 G. Olofsson, S. Nyholm, *Surface Potential Measurements*. In Proceedings, 2nd International Symposium "Protection Against Chemical Warfare Agents", Stockholm, Sweden, June 1986, pp. 430–431
6 K. B. Sipponen, *J Chromatogr*, 389, 87 (1987)
7 H. H. Willard, L. L. Merritt, J. A. Dean, F. A. Settle jr, *Instrumental Methods of Analysis*, Ed. 6, New York, 1981, p. 475
8 *Air monitoring as a means for verification of chemical disarmament*. C4, Part III, Helsinki, 1987, pp. 92–99
9 *Bruker Almanac* 1988, p. 49
10 J. J. Valdes et al, *Receptor-based biosensors*, Proceedings 3d International Symposium Protection Against Chemical Warfare Agents, Umea, Sweden, June 1989, p. 81
11 R. Taylor, *Receptor- and antibody-based biosensors*, see ref. 10, p. 83
12 Standard Operating Procedures for the verification of chemical disarmament. D.1, Helsinki, 1988, pp. 133–147
13 Air monitoring as a means for verification of chemical disarmament. C.2, Part I, Helsinki, 1985, pp. 129–148. Also A. Hesso, R. Kostiainen, *Tandem Mass Spectrometry*, see ref. 5, pp. 257–260
14 J. Franzen, M. Weiss and A. G. Loudon, *A Novel Sample Separat- ing Inlet for Chemical Agent Detection Using The MM-1 Mobile Mass Spectrometer*, see ref. 5, pp. 231–234
15 U. M. Lakkisto, *Retention Spectrometry – A new method for rapid and reliable detection of chemical warfare agents*, see ref. 5, pp. 245–250. *Retention Spectrometry*, see ref. 10, pp. 117–129
16 O. Kostiainen and R. Kostiainen, *Identification of CW agents by a three-channel GC coupled with ECD, ATD and ITD*, see ref. 5, p. 426
17 P. Kuronen, *Retention Indexes of Chemical Warfare Agents in Reversed-Phase High-Performance Liquid Chromatography*, see ref. 5, pp. 261–268
18 K. B. Sipponen, *An Enzymatic HPLC Detector System for the Analysis of Environmental Samples for Nerve Agents*, see ref. 5, pp. 269–273
19 D. G. Jones, *Anal. Chem.*, 57, 1057, 1207 (1985)
20 *The Diode Array Detector*, see ref. 8, pp. 49–65
21 E. Rahkamaa and M. Moilanen, NMR studies on some trichotecene mycotoxins, see ref. 5, pp. 275–283
22 Y. Margalit, G. Amitai and Y. Ashani, *Phosphorus and Sulpfur*, 3, 315 (1977).
23 Van den Berg, H. C. Beck and H. P. Benschop, *Bull. Environ. Contam. Toxicol*, 33, 505 (1984)
24 *Infrared Spectrometric Analysis*, see ref. 8, pp. 111–115
25 D. F. Gurka, M. Hiatt and R. Titus, *Anal. Chem.*, 56, 1102 (1984)
26 C. J. Wurrey, S. Bourne, R. D. Kleopfer, *Anal. Chem.*, 58, 482 (1986)
27 B. Odernheimer and G. Magin, *A Mobile MS-Detection System for Organic Pollutants*. In: Proc. of the 1-the International Symposium "Protection Against Chemical Warfare Agents", Stockholm, Sweden, 1983, p. 61
28 J. Franzen, M. Weiss and A. G. Loudon, *A novel sample separating inlet for chemical agent detection using the MM-1 mobile mass spectrometer*, see ref. 5, pp. 231–234
29 R. D. Plattner, G. A. Bennet and R. D. Stubblefield, *Identification of Aflatoxins by Quadrupole Mass Spectrometry*, J. Assoc. Anal. Chem., 67, 734 (1984)
30 Mats Ahlberg et al, *Chemical Analysis of Two Alleged Chemical Weapons Samples*, FOA-Report (Sweden) D-40117-C1, March 1984 (from: NBC Defence & Technology Int., April 1986, pp. 62-65)
31 *Technical evaluation of selected scientific methods for the verification of chemical disarmament*, Helsinki, 1984
32 Pia Bruce, *High volume air sampling with a radial sampler*, see ref. 5, pp. 207–282
33 J. Enquist, M. Kokko, M. Parjavinen, U. Lakkisto, *High resolution GC*, see ref. 11, pp. 207–182
34 A. Hesso, *High resolution selective ion monitoring with GC/MS*, see ref. 11, pp. 129–132, 146
35 Committee on Disarmament. CD/306, 10 August 1982
37 The Chemical Industry and the Projected Chemical Weapons Convention, Proc. of a SIPRI/Pugwash Conference, SIPRI C & BW Studies, vol. 5, part II, pp. 143–171 (1986)
38 A. Verweij, H. Boter and C. E. Degenhardt, *Science*, 204 (1979), p. 616

39 P. Kuronen, *High Performance Liquid Chromatographic Analysis*, In: "Systematic Identification of Selected Trichotecenes, Aflatoxins and Related Mycotoxins", Helsinki, 1986. Report B.5

40 K. B. Sipponen, *Detector for Organophosphorus Compounds in Liquid Chromatography Based on the Cholinsterase Inhibition Reaction*, J Chromatography, 389, 87 (1987)

41 E. Rahkamaa and M. Moilanen, see ref. 39, pp. 87–118, 181–186

42 See ref. 8, pp. 100–110

43 See ref. 8, pp. 111–115

The Chemical Weapons Convention: Particular Concerns of Developing Countries

Esmat Ezz

1. Proposed Chemical Weapons Convention

The International Community, as represented in the Paris Conference held in January 1989, has stressed the necessity of concluding at an early date a convention for a total ban on chemical weapons. It called on the Conference on Disarmament in Geneva to redouble its efforts, as a matter of urgency, to resolve expeditiously the remaining issues, and to conclude the Convention at the earliest date. The current period of global confidence-building, in particular between the two superpowers which had led the way to the Intermediate Nuclear Forces (INF) Treaty, and the acceptance by the Soviet Union of mandatory onsite challenge inspection within 48 hours as a verification measure in the CW Convention, should remove the major obstacles which have been delaying reaching an agreement.

The negotiations in Geneva have been going on for about 15 years; they have served to create common understanding of the necessary basic aspects of the problem and helped to remove many stumbling blocks on the way to achieve a universal ban on chemical weapons under effective international control. Lately, the discussions have been concentrating on the issues of verification, and in particular on "verification of non-production of chemical weapons in the civil chemical industry". In this particular aspect, the developing countries have a limited and/or primitive industrial and technological base which would make it easy to monitor and spot any attempt to violate the provisions of a convention. If these countries indulge in chemical weapons production, they would, in most cases, rely on the developed countries for many of their basic needs and that makes "verification of compliance of the developing countries" much easier. On the other hand, the technological and industrial capabilities, which constitute the capacity for chemical weapons production, are widely diffused within the industrial and economic structure of the developed countries, and their use in secret production of chemical weapons would be difficult to trace, unless intrusive verification measures are worked out. But these negotiations have never given sufficient consideration to the special concerns of the developing countries, and in particular to their security requirements.

The key point for the effectiveness of a chemical weapons convention is that all states adhere to it. To achieve global adherence, it is important to alleviate the worries of the developing countries and address their security concerns. If some countries do not join, the effectiveness of such a convention could be gravely harmed, as this would create islands of mistrust in the international system, and we could end up with a situation similar to that relating to the Non-proliferation Treaty (NPT), where a number of states did not join, went ahead with their military nuclear programmes, and are now threatening states which joined the NPT in good faith. This problem should be seriously considered during the negotiations for the proposed chemical weapons convention because the technology required for the production of chemical weaponry (CW) is within the reach of many countries. This could create a situation where a state does not join the convention, and could acquire CW, and be in a position to threaten or even use such weapons against a state party to the convention, and the world community might do nothing except, perhaps, condemn and deplore such action.

2. Security Needs

Peace and security are in the highest priority objectives of any state. Each country has a legitimate right to security and to engage in self-defence. It is evident that the majority of the developing countries have been diverting a good part of their meagre resources to meet the burdens of military expenditure, which they feel necessary to deal with the threats to their own security. Military build-ups would guarantee security only if they are of the quality and quantity that can deter the type of attack to which the country might be exposed. Such a build-up can also be effective, even in the absence of military balance between two adversaries, if the weaker side is capable of inflicting qualitative damage which the other side cannot tolerate.

Although the choice of weapons in armed conflicts is regulated by International Law and the other relevant treaties, we have been witnessing deliberate violations of these rules and regulations. Perhaps this is because, as J. P. Perry-Robinson once put it: [1]

> "...*when the national security is at stake, international treaties are bound to take a very second place to national self-interest, so much that arms-control or disarmament agreements are unlikely to have much practical influence: they will be observed only when that is expedient.*"

The spread of weapons of mass destruction has created a great deal of instability and destabilization in the world and is constituting a continuous threat to the security of the countries which do not possess such weapons. This is particularly true in conflict-ridden regions. Possession of biological weapons is prohibited by the Biological Weapons Convention of 1972. The Geneva Protocol of 1925 prohibits the use of chemical and biological weapons. As to nuclear weapons, there is no legal instrument which prohibits their use. True, the NPT has been designed to prevent the spread of nuclear weapons and ultimately to eradicate them from the world. But so far, the nuclear-weapon countries have not yet moved towards this objective, even though the non-nuclear-weapon signatory countries have been honouring their obligations. Moreover, some of the states which have not joined the NPT have gone ahead with their nuclear weapon programmes and developed their own nuclear weapons. This situation has created, in spite of the existence of the NPT, two categories of countries, nuclear-weapon countries and non-nuclear-weapon countries. As a result, countries which joined the NPT and honoured their obligations have not benefited from it, and even countries in some regions, such as the Middle East and South Africa, are now under the threat of nuclear weapons, and the world community is doing nothing positive to remedy this injustice.

There is as yet no international instrument which prohibits possession or production of chemical weapons. But it is envisaged that countries which join the proposed chemical weapons convention would agree to eschew chemical warfare and any intention of acquiring chemical weaponry. The question is how many chemical weapons can contribute to a country's security? And what are their real military value? Some argue that, if countries which possess the most sophisticated conventional and nuclear arsenals are not only maintaining their chemical weapons, but are also continuously modernizing them (for security requirements as they claim), therefore these weapons could be of real military value. This would at least make the idea of acquiring such weapons appealing to countries which do not have them, and at the same time do not have or could not have sophisticated conventional or nuclear weapons.

Some countries are tempted to *go chemical* to redress what they feel is an intolerable and unjust imbalance with neighbours which have the nuclear bomb. They argue: "what is the alternative to nuclear weapons for a country which cannot obtain such weapons, and is under the threat of these weapons by an adversary?". The question is whether chemical weapons could practically serve as a deterrent against a nuclear attack or could be used in retaliation after such an attack? No one can doubt that there are huge differences between the destructive power of nuclear weapons and chemical weapons. But some are concerned about a situation in which a country cannot acquire nuclear weapons but feels threatened by such weapons. If such a country has nothing to hand to face such a situation except chemical weapons, and it can acquire them, would it not be foolish if it did not resort to such an option to deter the threat of nuclear attack or even to retaliate after such an attack in order to inflict on the aggressor as much punishment as possible? Some countries even ask: "Why do you want to eliminate a weapon that both of us have or can have, while you do not want to give up a weapon that you possess, and I do not and cannot possess?"

3. The Value of a CW Convention

To quote Perry-Robinson again: [1]

> *"The value of the treaty lies in the size of benefits to be gained by joining the regime and the size of penalty for staying outside it."*

I would add that the real value of adhering to a treaty lies in the extent to which it could add to the overall security of the country concerned. Of course, the problem of security is multifaceted and the type of weapon, though decisive, is just one feature of a wider picture. Some countries may argue: are chemical weapons the only threat to our security? What about the other more serious weapons?

During the Paris Conference in 1989 many developing countries supported the global efforts to impose a total ban on chemical weapons, but they asked that the projected chemical weapons convention should include effective security guarantees for its members, not only against CW but against all weapons of mass destruction. The Arab States firmly united around a call for linking the elimination of CW with the elimination of nuclear weapons, and in particular the elimination of the nuclear capability of Israel. Egypt's Minister of Foreign Affairs said: [2]

> *"It would not be logical for the international community to permit to some countries in the most sensitive regions of the world the nuclear option without the least international control, while the same international community demands the total prohibition of chemical weapons. We consider that the progress in the field of the prohibition of chemical weapons is linked to the realization of a parallel prohibition on the level of nuclear weapons."*

Countries which possess nuclear weapons refused, however, to link the proposed ban on CW with the ban on nuclear weapons. This attitude brings to mind an analogy. Suppose I observe two opponents, one having only a knife and the other having a knife and a machine gun. Would I be justified in asking both to drop their knives, in order to establish peace between them, while ignoring the machine gun kept by one side? It thus became clear in Paris that many countries would adhere to the proposed CW convention only if their security concerns against all weapons of mass destruction are seriously taken into consideration.

4. Is There a Solution?

Perhaps the logical solution is to work out a practical and reliable system for international collective security under which all countries would feel safe and secure against the threat or use of any weapons of mass destruction. The question is whether this is politically feasible. Another solution, which may be more practical and could solve the particular concerns of countries in the areas where the nuclear threat exists, is to establish a zone free from weapons of mass destruction. There are sound grounds for supporting this proposal, at least in the Middle East. Both Egypt and Israel have asked in the United Nations that a nuclear-weapon-free zone be established in the Middle East. Yitzhak Shamir, the Israeli Prime Minister, proposed at the UNSSOD-3 on 7 June 1988, "a CW free zone to be established in the Middle East ... on the basis of arrangements to be arrived at freely among the states of the region". If the Israelis are sincere, the establishment of a zone free from all weapons of mass destruction under strict international control would thus remove a stumbling block to global adherence to the proposed CW convention.

5. Restrictions on Chemical Technology Transfer

It is generally accepted that certain controls should be imposed on import and export of chemicals and chemical technology to guard against CW proliferation. The difficulty is where to draw the line between what should be permitted and what should be controlled.

The subject of imposing restrictions and controls on the transfer of chemical technology, equipment and chemicals to the developing countries has been increasingly discussed and considered at various levels, starting with the Summit between Presidents Ronald Reagan and Mikhail Gorbachev down to meetings of non-governmental organizations. As a matter of fact, in one of those meetings organized by the UN Quaker Group held in April 1989 in Les Avants, Switzerland, a delegate to the Conference on Disarmament from a neutral European country mentioned that the objective of the projected CW Convention should be to control the chemical industry and chemical technology in order to prevent CW proliferation.

The developing countries are very much concerned about the emerging strategy of imposing excessive and unjustified controls on the transfer of chemical technology. The developing countries greatly need to promote their chemical industries to satisfy their increasing demand for chemicals for legitimate application to agriculture, industry and health services. They feel that part of their economic problems is due to reliance on imports for their needs, and this has been increasing their trade deficits. They felt that unjustified controls on chemical technology transfer could frustrate their efforts to promote industrial expansion and pose immense problems for their sagging economies. It is hoped that these particular worries can be taken care of by better co-ordination rather than by strict controls.

References

1 J. P. Perry Robinson, "Benefits and Costs of a Chemical Warfare Disarmament Regime". Conference on *Towards a Universal Chemical Weapons Convention*, Les Avants, 13–16 April 1989

2 Paris Conference on the Prohibition of Chemical Weapons, 7–11 January 1989. Public Plenary

The Chemical Weapons Convention: A Different View

Enrico Jacchia

The paper by Esmat Ezz is exactly what was needed to illustrate in an honest and clear way the general thinking of the political leaders of developing countries on the possession or the acquisition of a chemical weapons arsenal. At the outset, I want to state my belief that it would be a calamitous risk for the civilized world to let a number of developing countries (not all, of course, just a number) acquire a chemical arsenal. Therefore, I have no other choice than to oppose most of Ezz's arguments, while at the same time recognizing his moral and intellectual integrity.

There is, however, one point of broad agreement, namely when he writes that if some developing countries "indulge in chemical weapons production, they would rely on the developed countries for many of their basic needs, and that makes verification of compliance of the developing countries much easier". (Easier, I would say, not much easier). The merchants of death sell whatever sells. European industries have provided raw materials, equipment and know-how for the construction and operation of the Rabta plant in Libya. President Muammar Quadahafi says that it is a pharmaceutical plant. Webster, the Director of the US Central Intelligence Agency, maintains that it is intended for chemical toxic agents. I incline to believe Webster. But he also stated, on several recent occasions, that some European firms continue to supply, illegally, materials and equipment to the Libyan factory. And this is precisely the point. It would be relatively easy to verify the compliance of a given developing country if the producers of the developed countries, all without any exception, complied with the export control rules. We know that those rules are full of holes and that the international collaboration in this field is weak. And it will remain so until the majority of the civilized countries agree that the only serious way to stop the merchants of death – be it nuclear or chemical death – is the death penalty.

There is no doubt, as Ezz states, that "if some countries do not join, the effectiveness of the Convention of Chemical Weapons could be gravely harmed". But there is no doubt that some countries, for various good and bad reasons, will not join. What do we do? Do we abandon our efforts? Will we attempt to use intimidation to force compliance on those who are reluctant to join? Will we say – and it is the easier option – that if the Convention is not universal there will be no Convention? Surely, we can do better.

Ezz touches a very important point when he states "the question is whether chemical weapons could practically serve as a deterrent against a nuclear attack or in retaliation after such attack". To deal concretely with this question we cannot speak in general terms; we need to envisage some very precise, specific situations. If, for example, Saudi Arabia is threatened with a nuclear attack by the Soviet Union, a Saudi chemical arsenal would serve no useful purpose as a deterrent or in retaliation after such attack. The purpose of this example is to show that against a nuclear threat by one of the five military nuclear-weapon powers no developing or developed nation could rely on chemical weapons for deterrence or retaliation. In fact, the situation of some countries which are tempted to *go chemical* to redress what they feel is an intolerable imbalance *vis-à-vis* their neighbours who have the nuclear bomb, is a very special one: it concerns Israel and its Arab neighbours. So let us deal with it.

The Israeli situation points to an element of nuclear deterrence that is seldom discussed by ethical theorists, namely the problem of adversaries who regard each other as wholly outside any shared universe of moral obligation. In the case of the encounter of Western civilization with other civilizations and religions, moral constraints were progressively removed from the practice of warfare, save when the fear of retaliation came into play. The Nazis used the term *Tiermenschen* or subhuman beings, to describe some of their adversaries in the Second World War, while the Jews were described as insects and exterminated with Zyklon B. After that experience the Israelis are reputed to never discount any threat to exterminate them and, as a result, whether we like it or not, the atom bomb has become a fundamental component of Israeli strategy.

But to return to our theme: will a chemical arsenal be useful to Israel's neighbours? Probably not. The Israelis, according to different and reliable sources, have between 50 and 500 atomic bombs. They have a panoply of these weapons for specific potential use and, accordingly, they have a panoply of delivery means. This means they have a *force de frappe* strategy. No military professional would seriously say that against the enormous destructive power of such a *force de frappe* a certain amount of chemical weapons would serve a useful purpose.

A final observation on the *military value* of chemical weapons. After the recent progresses in the US-Soviet negotiations to eliminate, at least, the chemical weapons that their respective armies keep stored in Europe, the belief has spread that chemical weapons are in fact not militarily useful. That may be true for the superpowers. Analysis of the Iraq-Iran conflict increasingly provides evidence that this is not equally true for local conflicts between less sophisticated adversaries. I cannot go deeper into this subject here. Nevertheless, we should be careful in dismissing the military value of chemical weapons in situations different from those which prevail in the European theatre. This means that it is in the interest of all (and especially in the interest of the Third World countries) to stress the urgent need for a *universal convention* banning chemical weapons production.

Part IV

Nuclear Weapons Proliferation and Testing

The Non-Proliferation Treaty: Status of Implementation and the Threatening Developments

Jozef Goldblat

1. Introduction

The need to restrain the military threat of nuclear energy has been evident to many people from the early days of the atomic age. Indeed, the very first UN General Assembly resolution, of January 1946, called for the elimination of nuclear weapons from state arsenals. In the same year, the Government of the United States, which was the first to manufacture these weapons and to use them, proposed the establishment of an international authority to control all atomic energy activities. This proposal, known as the Baruch Plan, met with no success. In 1949 the Soviet Union also became a nuclear-weapon power, followed in 1952 by Great Britain, in 1960 by France, and in 1964 by China. Subsequently, however, the realization that proliferation of nuclear weapons would pose a danger to world security led to the development of a non-proliferation regime which encompasses various restrictive rules, as well as specialized control institutions both national and international. Among the latter, the International Atomic Energy Agency (IAEA) fulfills an essential practical role, but the pivotal place in the regime belongs to the Treaty on the Non-proliferation of Nuclear Weapons (NPT), signed in 1968. The NPT is a unique document in the sense that it prohibits the possession by an overwhelming majority of states of the most destructive weapons yet invented, while tolerating the retention of the same weapons by a handful of nations. But the NPT is not an end in itself: the declared aim of the parties is to use it as a transitional measure to clear the way towards nuclear disarmament.

In spite of the inequality of treaty rights and obligations of the nuclear and non-nuclear weapon parties, the NPT, in force since 1970, has attracted a record number of adherents for an arms control agreement – 140. These include three nuclear-weapon powers – Great Britain, the United States and the Soviet Union – as well as almost all highly developed, industrialized and militarily significant non-nuclear-weapon states. France, a nuclear-weapon power which has not signed the NPT, has a declared policy of behaving like a state party to it. China, the fifth nuclear-weapon power, has given solemn assurances that it would not help other states to acquire nuclear weapons.

In the course of the past two decades the non-proliferation regime has been considerably strengthened. No material breaches of the NPT have been recorded, and no withdrawals have been notified. However, the non-proliferation regime is also experiencing a few disquieting trends. This is due to the planned acquisition of nuclear-powered submarines by non-nuclear-weapon states, the growing trade in nuclear-capable missiles, and the emergence of new suppliers of nuclear hardware and services.

2. Status of Implementation

2.1 Non-Transfer and Non-Acquisition of Nuclear Weapons

The essential non-proliferation undertakings are contained in the first two Articles of the NPT. Article I places the nuclear-weapon states under the obligation not to transfer "to any recipient whatsoever" nuclear weapons or other nuclear explosive devices or control over them, and not in any way to "assist, encourage, or induce" any non-nuclear-weapon state to manufacture or acquire such weapons or devices. Article II pledges the non-nuclear-weapon states not to receive nuclear weapons or other nuclear explosive devices or control over them, as well as not to manufacture them or receive assistance in their manufacture.

The implementation of the first part of Article I prohibiting transfer of nuclear explosive devices, though unverifiable, has not given rise to formal complaints. The extent to which the second part of Article I has been observed – the part prohibiting the provision by the nuclear-weapon states of assistance in the manufacture of nuclear weapons – has lent itself to controversy. Since there exists a significant overlap between the technologies of civilian nuclear energy and those useful for military explosive purposes, and since nuclear material and technology nominally destined for power programmes have been exported by NPT parties to countries which have not formally forgone nuclear weapons, it is argued by some that the obligation not in "any way" to assist non-nuclear-weapon states to manufacture nuclear explosive devices has not been fully complied with.

As regards Article II, there is no evidence that any non-nuclear-weapon party to the NPT has clandestinely manufactured or otherwise acquired nuclear explosive devices. If any one of them has designed a nuclear weapon or even developed its non-nuclear components, these activities would be difficult to detect. Should such a state ever decide actually to produce a nuclear weapon, it would need the requisite quantity of weapon-grade fissile material. The availability of this material is, therefore, of crucial significance; hence the importance of safeguards to prevent its diversion from peaceful to military uses.

2.2 Nuclear Safeguards and Protection of Nuclear Material

The safeguards requirement under Article III constitutes the verification element of the NPT. Safeguards should enable detection of diversion of significant quantities of nuclear material from peaceful activities to the manufacture of nuclear explosive devices, as well as deterrence of diversion by creating the risk of timely detection. No such diversion has as yet been reported by the IAEA, although on several occasions the Agency has been hindered in its inspection activities.

The Treaty requires safeguards to be implemented in such a manner as to avoid hampering the economic or technological development of the countries party to it, or international cooperation in the field of peaceful nuclear activities. This requirement seems to have been met, although there have been some complaints that controls complicate the production process, or are a burden for enterprises in terms of cost and of protection of industrial secrets. More controversial is the clause setting forth the conditions for nuclear trade with non-nuclear-weapon countries. This clause has been applied in a way that has sometimes benefited non-parties more than parties. For whereas parties are subject to NPT safeguards covering all their peaceful nuclear activities, the nuclear activities of non-parties are covered only partially by safeguards of the pre-NPT order, which apply exclusively to imported items – individual installations or material – while part of the nuclear fuel cycle may remain unsafeguarded. Many suppliers concerned about the dangers of nuclear proliferation inherent in the distinction between imported and domestic technology have sought to impose on non-parties full-scope safeguards, as extensive as NPT-type safeguards. A few suppliers, however, are reluctant to modify radically their export conditions.

An important step towards reducing the risks of diversion of nuclear material to non-peaceful purposes was made in 1987, with the entry into force of the 1980 Convention on the Physical Protection of Nuclear Material. The provisions of the Convention oblige the parties to ensure that, during international transport across their territory or on ships or aircraft under their jurisdiction, nuclear material for peaceful purposes, as categorized in a special annex (plutonium, uranium-235, uranium-233 and irradiated fuel), is protected at the agreed level. Furthermore, the parties undertake not to export or import nuclear material or allow its transit through their territory unless they have received assurances that this material will be protected during international transport in accordance with the levels of protection determined by the Convention. The parties to the Convention agree to share information on missing nuclear material to facilitate recovery operations. Robbery, embezzlement or extortion in relation to nuclear material, and acts without lawful authority involving nuclear material which cause or are likely to cause death or serious injury to any person or substantial damage to property, are to be treated as punishable offences.

Towards the end of 1987, alarm was raised in West Germany and Belgium because of alleged illegal cross-border transportation of canisters with nuclear wastes. It was asserted that at least some canisters were falsely labelled and actually contained fissionable material destined for Pakistan and Libya. No evidence was supplied to support this allegation [1]. It seems, none the less, that some serious irregularities in the transportation of radioactive substances did take place. The members of the European Community have not yet ratified the Physical Protection Convention, even though shipment of nuclear material in the territories of the Community is very intensive.

2.3 Peaceful Uses of Nuclear Energy

Article IV of the NPT reaffirms the right of parties to develop nuclear energy for peaceful purposes in conformity with Articles I and II of the Treaty, and obligates those parties in a position to do so to contribute to such efforts in non-nuclear-weapon states. The implementation of Article IV has been affected to a great extent by the world-wide slowdown in the growth of civilian nuclear power owing to environmental, safety and economic factors, which include a weak increase in electricity demand, high initial investment costs, and shortage of capital. In addition, restrictions have been imposed on nuclear supplies ever since India, a non-party to the NPT, took advantage of the lenient pre-NPT safeguards applied to its nuclear activities to explode a nuclear device. In 1977, a group of nuclear suppliers, the so-called London Club, drew up a list of materials, equipment and technology which should "trigger" IAEA safeguards when exported to non-nuclear-weapon states not party to the NPT (the parties having already accepted the requirement for safeguards on all their nuclear activities). The Guidelines for Nuclear Transfers agreed by the London Club require the recipients of the trigger-list items to provide effective physical protection for these items, and to pledge not to use them for the manufacture of nuclear explosives. The safeguards requirements apply to any *replicated* facility, that is, of the same type as the imported facility but constructed indigenously within a specified period. Retransfers of trigger-list items are to be subject to the same conditions as those attached to the original transfer. In the event of the diversion of materials or a violation of the supplier/recipient understandings, the members of the London Club should consult promptly on possible common action. Moreover, *restraint* is recommended in the transfer of sensitive facilities, such as uranium enrichment and plutonium reprocessing plants. Significantly, this restraint is to be exercised not only with respect to non-parties, but also with respect to parties to the NPT. In 1978 the United States unilaterally set even sterner restrictions on nuclear supplies by adopting the Nuclear Non-proliferation Act (NNPA).

Spokesmen of many countries, especially from the Third World, have criticized the restrictive measures taken by the suppliers as an infringement of the right to nuclear supplies implied in NPT Article IV. Their argument is that once governments have accepted the safeguards provided for in Article III, no further limitation should be placed on peaceful nuclear programmes. The Committee on Assurances of Supply which was set up by the IAEA to consider and advise on "ways and means in which supplies of nuclear material, equipment and technology and fuel cycle services could be assured on a more predictable and longterm basis in accordance with mutually acceptable considerations on non-proliferation" has not, as yet, produced agreed principles of international cooperation. Also the UN Conference on the Peaceful Uses of Nuclear Energy, which met in Geneva in March-April 1987, failed to work out such principles, mainly because non-parties to the NPT refused to take account of nuclear-weapon proliferation concerns related to supplies of nuclear material and equipment.

2.4 Peaseful Nuclear Explosions

Under Article V of the NPT, the potential benefits of peaceful applications of nuclear explosions are to be made available by the nuclear-weapon parties to non-nuclear-weapon parties under appropriate international observation. This promise was made in exchange for the renunciation by the latter states of the right to conduct any nuclear explosions, because there is no way to assure that a nuclear explosion has no military function.

However, there is considerable scepticism about the technical feasibility, economic viability and political acceptability of nuclear explosions for peaceful purposes. It is now recognized that conventional explosives can achieve equivalent results without the environmental and health risks accompanying nuclear detonations. As a matter of fact, there has so far been no project in which the supply of peaceful nuclear explosion services to a non-nuclear-weapon state was necessary. The United States terminated its peaceful nuclear explosion programme in 1977, and the only country which still experiments with such explosions – although it says it is willing to cease them under an overall nuclear test ban agreement – is the Soviet Union. The prevailing opinion, at least among the parties to the NPT, seems to be that peaceful uses of nuclear explosions entail more hazards than benefits. By tacit agreement, therefore, the practical implementation of this provision has been kept in abeyance.

2.5 Disarmament Obligations

The obligations under Article VI are generally considered to be of particular consequence. For in signing the NPT the parties agreed that the self-imposed arms denial of one side – the non-nuclear-weapon states – was to be matched, ultimately, by corresponding acts of the other side – the nuclear-weapon powers. They have, therefore, undertaken to pursue negotiations in *good faith* to halt the nuclear arms race *at an early date* and to bring about nuclear disarmament. In fact, the NPT is the only existing international document under which the major nuclear powers are legally committed to nuclear disarmament. However, the nuclear arms control agreements concluded in the 1970s – the 1972 SALT Interim Agreement and the 1979 SALT Treaty – were of low disarmament value, as they merely regulated the US-Soviet competition at a high level of armaments. Besides, these agreements are no longer in force, and the validity of the 1972 ABM Treaty restricting ballistic missile defences is being jeopardized by developments in military space technology. The first meaningful measure of nuclear disarmament was taken in 1987 with the signing of the US-Soviet Treaty eliminating all ground-launched missiles with a range of 500 to 5500 km. A 50 per cent reduction of strategic missiles having a range in excess of 5500 km is under negotiation.

Whereas the sheer numbers of nuclear weapons seems to be on the decline, no effective measure has been taken so far to restrain the qualitative improvement of these weapons. In this respect a comprehensive ban on nuclear weapon testing would be essential. This is why the preamble to the NPT reiterated the determination of the parties to the 1963 Partial Test Ban Treaty (PTBT) to achieve the discontinuation of all test explosions of nuclear weapons *for all time*. Indeed, without tests the development of new designs of nuclear weapons by the nuclear weapons powers would be made practically impossible, while the modification of existing weapon designs would be seriously constrained. But the trilateral British-US-Soviet talks on a comprehensive test ban treaty have remained suspended since 1980, and the multilateral talks on the same subject held at the Conference on Disarmament (CD) have been stalemated in a dispute over the terms of reference of the Conference working committee.

2.6 Nuclear-Weapon-Free Zones

Article VII of the NPT affirms the right of states to conclude regional treaties in order to assure the total absence of nuclear weapons in their respective territories. Two such treaties covering large populated areas have so far been concluded: the 1967 Treaty of Tlatelolco, prohibiting nuclear weapons in Latin America, and the 1985 Treaty of Rarotonga, setting up a nuclear-free zone in the South Pacific. As an underpinning of the Treaty of Rarotonga, the Parliament of New Zealand adopted in 1987 an act establishing the New Zealand Nuclear-Free Zone which comprises all land and waters within the territorial limits of New Zealand, as well as the airspace above these areas. The Act states that the Prime Minister may grant approval for the entry of foreign warships into the internal waters of New Zealand only if he is satisfied that the warships will not be carrying any nuclear explosive device upon their entry into these waters. Similarly, approval for the landing in New Zealand by foreign military aircraft may be granted only by the Prime Minister if he is satisfied that the aircraft will not be carrying any nuclear explosive device when it lands. Entry into the internal waters of New Zealand by any ship whose propulsion is wholly or partly dependent on nuclear power is prohibited [2].

In 1988, at the initiative of the opposition Social Democratic Party, the Parliament of Denmark passed a resolution requesting the Government to notify all visiting warships that they must not carry nuclear arms into Danish ports. From the formal point of view, the resolution merely reiterated the official Danish policy, which had been proclaimed more than three decades earlier, namely, that the introduction of nuclear weapons to the country is prohibited during peacetime. In practice, however, the resolution signified a rejection of the policy of "neither confirming nor denying" the presence of nuclear weapons, which has been so far strictly adhered to by the navies of all the nuclear weapon powers. Eventually, however, under pressure exercised within NATO, mainly by the United States and Great Britain, Denmark agreed, as a compromise, to adopt the Norwegian formula. Norway, which has also unilaterally declared its territory to be free of nuclear weapons in peacetime, proceeds on the assumption that this declaration is respected by the visiting foreign ships or aircraft and does not seek specific assurances [3]. Several other countries as well, including members of the military alliances, have formally prohibited (as have Japan, Iceland and Spain) or have contemplated prohibiting (as has the Philippines [4]) foreign ships or aircraft to enter their territories with nuclear weapons aboard. None, however, has so far tried to enforce this prohibition.

In Sweden, the ruling Social Democratic Party decided, at its 1987 congress, that efforts should be made to make the nuclear powers forgo the practice of not giving information regarding the presence of nuclear weapons on their ships. It was resolved that, should the nuclear powers decline to give up this practice, the rules for military visits would be tightened: the powers in question would be requested to make an explicit statement that nuclear weapons were not entering Swedish territory, including the airspace. The visits would be refused if no such information were provided [5]. This policy was confirmed by the Swedish Prime Minister in his speech made at the 1988 Third UN Special Session on Disarmament.

3. The Threatening Developments

3.1 Naval Propulsion

In 1988 it became known that India had leased a nuclear-powered submarine from the Soviet Union [6] and that the submarine was equipped with cruise missiles [7]. The lease was apparently intended to lead to the purchase of several submarines by India and/or help it to develop an indigenously-designed nuclear-propulsion system. The conditions of the leasing arrangement have not been made public. However, since Soviet submarines use highly enriched uranium which can also be used to produce nuclear weapons, and since India refuses to forsake the nuclear weapon option, the compatibility of the Soviet-India deal with the goal of non-proliferation is open to question.

Another instance of nuclear-propulsion proliferation is the planned acquisition by Canada of a fleet of nuclear-powered submarines equipped with conventionally armed torpedoes8. However, as distinct from India, Canada is party to the NPT and has accepted full-scope IAEA safeguards. Consequently, the concerns regarding this acquisition are not of the same order as in the case of India. As a matter of fact, under the NPT, non-nuclear-weapon states are prohibited only from using nuclear materials for explosive purposes; the use of such materials for naval propulsion is not prohibited. Paragraph 14 of the "Structure and Content of Agreements between the Agency and States", required in connection with the NPT, provides for a special arrangement for withdrawing nuclear material from IAEA safeguards, so that it can be used in non-proscribed military activities [9]. The arrangement between the state in question and the IAEA should identify the circumstances during which safeguards would not be applied. The state would have to make it clear that the unsafeguarded material (the quantity and composition of which must be known to the IAEA) would not be used for the production of nuclear weapons or other explosive devices. According to the authoritative interpretation, based on the negotiating history, the exemption from safeguards is to be strictly limited to the material in the propulsion reactors and should not include other stages of the nuclear fuel cycle; safeguards must again apply as soon as the nuclear material is re-introduced into a peaceful nuclear activity for reprocessing or for other, inherently non-military industrial treatment [10]. If Canada comes into possession of nuclear-powered submarines, it may avail itself of the exemption provision referred to above. This may not affect its commitment to the cause of non-proliferation of nuclear weapons, but would set an unfortunate precedent for the non-application of nuclear safeguards by the parties to the NPT [11].

Among other states known to plan the acquisition of nuclear-powered submarines are Argentina and Brazil, both holdouts from the NPT [12].

3.2 Dual-Capable Missiles

One recommendation frequently made for the strengthening of the non-proliferation regime has been to complement the existing restraints on supplies of nuclear material and equipment by restraints on supplies of dual-capable weapon systems, that is, systems capable of delivering both a conventional and a nuclear weapon. This recommendation was partly put into practice in April 1987 when seven governments, those of Great Britain, Canada, France, West Germany, Italy, Japan and the United States, adopted identical guidelines to control the exports of equipment and technology which "could make a contribution" to missile systems capable of delivering a nuclear weapon.

The guidelines are accompanied by an annex of items to be controlled, and the items fall into two categories. Item 1 of category I covers complete rocket systems (including ballistic missile systems, space-launched vehicles and sounding rockets) and unmanned air vehicle systems (including cruise missile systems, target drones and reconnaissance drones) capable of delivering at least a 500-kg payload within a range of at least 300 km, as well as the specially designed production facilities for these systems. Item 2 of the same category covers complete sub-systems usable in item 1 systems, as well as the specially designed production facilities and production equipment therefor. All other items (3–18) in the annex are category II items. Particular restraint is to be exercised in the consideration of category I item transfers because of their great sensitivity, and there is a presumption of denial with respect to such transfers. Until further notice, transfer of category I production facilities will not be authorized. Category II items are not on a "denial" list. Almost all items in this category have uses other than for projects of nuclear proliferation concern and their transfers will be considered on a case-by-case basis. Nuclear proliferation occupies a prominent place among the factors which must be taken into account in the evaluation of export applications.

The missile export control guidelines constitute an important initiative, in so far as they can make it more difficult, and perhaps more expensive, for countries to acquire a nuclear weapon delivery capability. However, the control regime is focused on large missiles and rockets; it is not designed to constrain more sophisticated forces. It ignores such an important and relatively easily available nuclear delivery vehicle as aircraft. Moreover, the restrictions have come somewhat late. Companies from West Germany, France and Italy have been collaborating for some time with Third World missile producers [13], and both the superpowers have provided different types of missile to several countries. In particular, Iraq, Iran, Libya, Syria and South Yemen are now in possession of Soviet-made missiles, some of which have been modified to reach a range of several hundred kilometres14. Negotiations are said to have been held for the sale of a new 600 km-range Chinese missile to Syria [15].

India has greatly developed its nuclear weapons delivery capability in addition to the aircraft it possesses. In particular, it has successfully tested the Prithvi short-range surface-to-surface rocket. Variously-described as having ranges of 150 and 240 kilometres, computer-guided and able to carry a payload of one tonne, the Prithvi has been favourably compared to the US Lance missile currently deployed with NATO forces in Europe. It is intended to become the basis for a family of missiles and is scheduled for mass-production in the 1990s; another missile under development in India is expected to have a range of 2400 km. The Indian space programme with its obvious military potential is also going ahead. India has already put a satellite into orbit with an indigenously-produced rocket. However, in this area serious and costly engineering setbacks have occurred suggesting that the development of Indian ballistic missiles of intercontinental range is still far away [16]. Israel has demonstrated that it possesses a rocket powerful enough to launch a satellite into space [17], and has reportedly deployed intermediate-range (over 1000 kilometres) ballistic missiles in the Negev Desert [18].

Pakistan has successfully test-fired surface-to-surface rockets capable of carrying a payload of more than 500 kg. The rockets are said to have been developed indigenously, though press reports allege that this has been done with the help of China [19].

Argentina is developing – in cooperation with Iraq and Egypt – a medium-range (800–950 km) two-stage rocket with a payload of some 350 km [20], whereas Brazil is known to manufacture and export a wide variety of rockets, some of which are capable of delivering nuclear weapon payloads [21]. A major deal was the purchase by Saudi Arabia of Chinese surface-to-surface ballistic missiles having a range of about 300 km, and designed to deliver nuclear warheads [22].

Some of the recipient countries may be many years away from a nuclear weapon manufacturing capability, but if they decided to go nuclear they would certainly adapt for nuclear delivery those missiles they already possessed. It is, therefore, widely recognized that to be more effective in reducing the risk of nuclear weapon proliferation, especially among non-parties to the NPT, the seven-nation missile technology control regime would have to be subscribed to by all suppliers of missiles, including of course the Soviet Union and China; the regime itself would have to be made sterner and its scope would need broadening. Exploratory talks on this subject have already been held between the United States and the Soviet Union [23].

3.3 The Emergent Nuclear Suppliers

The world nuclear market is dominated by a handful of industrial nations. However, the pattern of supplies is gradually changing as former buyers are acquiring the capability of developing their own nuclear technology and become themselves sellers of nuclear hardware and services. The new suppliers are mostly from the Third World. Their share in the global nuclear trade is still very modest. They are not in a position to provide modern sophisticated equipment, but they may offer an attractive alternative for those countries which shun the restrictive policies of the traditional suppliers.

So far, there has been no significant damage done to the non-proliferation regime by the emergent suppliers, because most transactions are internationally safeguarded. But the new-comers – among whom the most active are China, Argentina, Brazil and India – may decide to be less demanding as regards the application of safeguards; they are not bound by the 1977 London Guidelines for Nuclear Transfers. As a consequence, surveillance of nuclear development, especially in non-NPT countries, could become even more difficult. Particularly destabilizing would be an uncontrolled trade in sensitive items, including nuclear spent fuel reprocessing and uranium enrichment technologies, for it could considerably weaken the non-proliferation regime. A dialogue would, therefore, be desirable between the emerging and established suppliers with a view to working out generally acceptable rules [24].

4. Conclusions

In spite of certain reverses in the field of nuclear export control, the nuclear non-proliferation regime has proved to be fairly robust. There is a good chance that the next, 1990, NPT Review Conference, will reaffirm the validity of, and the support for, the NPT, and that the 1995 Conference, which is to decide the Treaty's future, will extend the duration of the NPT for another lengthy period.

The danger that a nuclear *threshold* country may openly join the club of established nuclear-weapon states continues to exist, but this danger could be greatly reduced if the regional disputes and conflicts feeding nuclear ambitions were settled. For non-proliferation has become a norm of international behaviour which cannot be easily defied. The ultimate solution to the problem of non-proliferation will, of course, be found only in a world in which the possession of nuclear weapons is universally recognized as both unnecessary and unacceptable. This goal is still remote. To bring it nearer, the process of nuclear arms reduction and elimination, initiated with the 1987 INF Treaty, must continue.

References

1 IAEA Press Release PR 88/3; *Süddeutsche Zeitung*, 26 February 1988

2 New Zealand Nuclear Free Zone, Disarmament and Arms Control Act 1987

3 *Berlingske Tidende* (Copenhagen), 15 April 1988; *Svenska Dagbladet* Stockholm) 20, 22, 23 and 24 April 1988; and *Le Monde*, 9 June 1988

4 *New York Times*, 7 June 1988

5 Swedish Social Democratic Party Congress Protocol MO-103-gh-0010, September 1987

6 *Pravda*, 6 January 1988

7 *Le Monde*, 11 February 1988

8 Dept. of National Defence, *Challenge and Commitment: A Defence Policy for Canada* (Canadian Ministry of Supplies and Services, Ottawa, 1987)

9 IAEA document INFCURC/153

10 D. Fischer and P. Szasz, ed. J. Goldblat, SIPRI, Safeguarding the Atom: A Critical Appraisal (Taylor & Francis, London, 1985); H. Blix "Aspects juridiques des garanties de l'Agence Internationale de l'energie atomique', *Annuaire francais du droit international*, vol. 29, 1983

11 For a more extensive treatment of this subject, see: B. Saunders and J. Simpson, "*Nuclear Submarines and Non-Proliferation: Cause for Concern*". Programme for Promoting Nuclear Non-Proliferation, Occasional Paper Two, 1988; M.-F. Desjardins, and T. Rauf, "*Opening Pandora's Box? Nuclear-Powered Submarines and the Spread of Nuclear Weapons*". The Canadian Centre for Arms Control and Disarmament, Aurora Papers 8, 1988

12 *Jane's Defence Weekly*, 23 July 1988

13 *Financial Times*, 8 June 1988

14 *Le Monde*, 8 April 1988; and *International Herald Tribune*, 2 May 1988

15 *International Herald Tribune*, 23 January 1988

16 *Neue Zürcher Zeitung*, 30 March 1988; *Aviation Week and Space Technology*, 26 October 1988; and *International Herald Tribune*, 4 April 1989

17 *International Herald Tribune*, 20 September 1988

18 L. S. Spector, "*Going Nuclear*". Ballinger, Cambridge, Mass., 1987; and *International Defence Review*, 21 July 1987

19 *International Herald Tribune*, 25 May 1988 and 6 February 1989

20 *Frankfurter Allgemeine Zeitung*, 22 December 1987; *International Herald Tribune*, 20 September 1988; *The Guardian*, 21 September 1988

21 A. Karp, "The frantic Third World quest for ballistic missiles", *Bulletin of the Atomic Scientists*, June 1988

22 *International Herald Tribune*, 19–20 March 1988; *Le Monde*, 7 April 1988

23 *New York Times*, 27 September 1988

24 For detailed proposals to this effect, see L. A. Dunn, "Nonproliferation The Next Steps", *Arms Control Today*, Washington DC, November 1987

Beyond Non-Proliferation

Bhalchandra Udgaonkar

1. Some History

The first atomic weapons were exploded over Hiroshima and Nagasaki in August 1945, six weeks after the signing of the United Nations Charter. In its very first resolution, arising from the initiative of the five permanent members of the Security Council and Canada, adopted in January 1946, the General Assembly of the United Nations unanimously decided to establish an Atomic Energy Commission which was to make *specific proposals for the elimination of atomic weapons from national armaments*. The Commission could not make any headway because of the irreconcilable differences between the superpowers. What survived its deliberations and related discussions at the United Nations and elsewhere, at the hands of the big powers (which one by one developed their own nuclear weapon capabilities in the meantime) was the idea of non-proliferation, decoupled from the disarmament dimension, in due course leading to the Treaty on Non-Proliferation of Nuclear Weapons (NPT) of 1968 [1]. While much is made of the fact that the NPT has 140 parties as of May 1989, one should note that the non-signatories include not only two nuclear-weapon powers, namely China and France, but also several threshold powers, and that 45 per cent of the world's population is in states which have not accepted the NPT. Even among some of the signatory states there have been serious discussions of going nuclear after 1995, if the present nuclear-weapon states are allowed to retain a minimum deterrence.

In 1961, the same year that Ireland presented a draft resolution to the General Assembly that was the first of a series that would lead to NPT, the non-aligned nations also moved a resolution in the General Assembly declaring that the use and threat of use of nuclear weapons would be a direct violation of the Charter of United Nations, and that any state using such weapons would be considered as acting contrary to the laws of humanity and as committing a crime against mankind and civilization. It demanded the total prohibition of the use of nuclear weapons. The resolution was then passed, with 55 (including all the nonaligned countries) voting in favour, 20 voting against (mostly NATO countries and their allies) and 26 (including the Socialist-bloc countries) abstaining. In more recent years, while more than 136 member nations of the UN have voted repeatedly calling for a ban on the use and the threat to use of nuclear weapons, the resolution has been persistently opposed by the 15 NATO countries and two of their allies. Such persistent opposition underlines the fact that the NATO countries, even if they may have signed the NPT, consider the nuclear weapons to be legitimate weapons. It also means that they do not take seriously the reminder in the preamble of the NPT, that in accordance with the Charter of the United Nations, states must refrain in their international relations from the threat or use of force against the territorial integrity or political independence of any State. All this is not calculated to inspire confidence in the NPT regime.

The approach of the non-aligned nations has been from the very beginning strongly against the legitimacy of nuclear weapons, as instruments of policy, and correspondingly against any discrimination between the nuclear weapons states and non-weapon states [2]. In the matter of non-proliferation, the second non-aligned summit at Cairo (1964) called upon the great powers to abstain from all policies conducive to the dissemination of nuclear weapons among states that did not possess them, and at the same time to agree on measures providing for gradual liquidation of the existing stockpiles of nuclear weapons. The conference also took an unambiguous stand against the deployment of nuclear weapons on the territories of non-nuclear states as one step towards gradual liquidation of the stockpiles and towards denuclearization.

In 1965 India and seven other nations presented a resolution to the General Assembly of the United Nations on A Treaty to Prevent the Proliferation of Nuclear Weapons, which called for negotiation of an international treaty to prevent the proliferation of nuclear weapons based on certain principles [3]. These included the following:

- The treaty should be void of any loopholes which might permit nuclear or non-nuclear powers to proliferate, directly or indirectly, nuclear weapons in any form;
- The treaty should embody an acceptable balance of mutual responsibilities and obligations of the nuclear and non-nuclear powers;
- The treaty should be a step towards the achievement of general and complete disarmament and, more particularly, nuclear disarmament;
- There should be acceptable and workable provisions to ensure the effectiveness of the treaty.

This resolution was passed by the General Assembly with an overwhelming majority which included Great Britain, the United States and the Soviet Union. The NPT which eventually emerged, however, did not embody the above concerns and principles. The objective of elimination of nuclear weapons almost disappeared from the international agenda for the next twenty years, thereby blocking the repeated reminders from the non-aligned movement and the peace movements. The superpowers followed a path of mutually agreed armaments build-up rather than one of mutually agreed phased arms reduction, showing scant respect for their obligations under Article VI of the NPT. And the other three nuclear powers also built up their own nuclear arsenals as they saw fit.

2. The Present Situation

Vertical proliferation has gone on apace during the last two decades, unaffected by the NPT, and the number of nuclear warheads in the world has increased from less than 10 000 at the time when the NPT was signed in 1968, to about 60 000 today, with greater and greater sophistication in weapons and delivery systems thrown in. In relation to the euphoria unleashed by the INF Treaty and the (so far unfulfilled) prospect of 50 per cent cuts in strategic weapons, one must note that even if 85 per cent of the present stockpile of nuclear warheads were to be dismantled, the world would still have more nuclear warheads than existed at the time of the signing of the NPT in 1968.

Testing of nuclear weapons, and improvements in weapons, have continued unabated. There has been no progress towards a Comprehensive Test Ban Treaty which was the declared intention of the signatories when the Partial Test Ban Treaty was signed in 1963, an intention which was also recalled in the preamble to the NPT five years later. Even the Threshold Test Ban Treaty between the superpowers (1974), and the similar agreement signed between them in 1976 to limit the yield of peaceful nuclear explosions, have not been ratified.

One may remark here that signing a threshold test ban treaty at this stage will be counterproductive. It will clearly imply a fresh legitimization of the nuclear arms race, including the development of new nuclear weapons.

The Pugwash Council, in the Statement issued by it after the Annual Conference at Dagomys in September 1988, ruefully noted: "There still is more talk about disarmament than there is action, and deployments of new weapons systems are proceeding at a rapid rate" [4].

3. Horizontal Proliferation Ignored

The number of declared nuclear-weapon states has long remained static at five. There has, however, been continued expansion, in both qualitative and quantitative terms, of the nuclear weapons storage, command, control and communications facilities that constitute the nuclear-war-fighting infrastructure, which now spread over at least 62 nations, principalities and territories [5]. There has also been an increase in the number and capabilities of nuclear-weapons-carrying submarines and other naval vessels in all oceans of the world, with the possible exception of the Antarctic. The horizontal proliferation that this implies, in terms of the number of independent, semi- and quasi-independent decision-makers *within* the national structures of the five individual nuclear weapons powers, has not received as much attention in discussions of the NPT, as the possible horizontal proliferation to non-nuclear-weapon states. Whatever pretence one may have of a cen- tralized political control of such a vast world-wide complex technological apparatus of nuclear weaponry in peacetime, as Paul Bracken has vividly pointed out, in a time of crisis, the nuclear command systems in the nuclear weapons states, including the superpowers, are likely to pass rapidly from political to highly fragmented military control [6]. Jasjit Singh has emphasized that the delegation of authority to such decision-makers is likely to increase in future as reaction-response times reduce, thus further increasing the number of independent decision-makers; and, further, that during times of high alert the decision-making was very likely to pass into the hands of so-called non-nuclear weapon allies too [7]. *The real horizontal proliferation* has thus been within the nuclear weapon powers, to non-nuclear weapon states and territories, and also to crypto-nuclear weapon powers, whose nuclear weapon programmes have been aided, abetted and connived at by the nuclear-weapon powers and their allies, thereby violating Article I of the NPT [8].

Discussions of horizontal proliferation are usually confined to the possible acquisition of nuclear weapons (or capability) by Third World countries. Not only is similar concern not visible regarding possible acquisition of nuclear weapons capability (and arsenals) by South Africa and Israel, but one finds hardly any discussion of the nuclear capabilities of Japan and many European countries, and of the nuclear capabilities of corporations, and what these imply for the stability of the NPT regime. Is this because the ethnocentric advocates of NPT are not worried about horizontal proliferation among their allies, and are concerned only about a possible spread to Third World countries which are not part of their alliance systems?

David Fischer made an interesting argument while discussing the nuclear weapons capability/arsenal of Israel [9]:

> "*... in all circumstances it is vital that the USA should continue to persuade Israel to refrain from openly proclaiming her nuclear capability. An Israeli nuclear test or public declaration that she was a nuclear weapon state would send a destabilizing shock-wave throughout the Middle East which could also have damaging consequences for the 1990 and 1995 Conferences and for the future of the NPT. It would face the USA with a painful dilemma. If the USA failed to react with sufficient vigour, her own non-proliferation policy would suffer a severe loss of coherence and credibility.*"

This argument hardly needs a comment.

We may thus conclude that global strategic and/or commercial considerations of nuclear-weapon states and their allies, and of their corporations, have often had precedence over the declared non-proliferation concerns of the big powers.

4. Doctrines and Attitudes

In the context of these developments, one has to examine the legitimizing role played by pernicious concepts and doctrines such as nuclear deterrence, flexible response, and limited nuclear war, which are logically inconsistent with a non-proliferation policy (unless it is merely aimed at keeping down the newly industrializing nations), and yet have not been given up by certain vociferous advocates of the NPT. Correspondingly, many of the so-called non-nuclear-weapon countries, which are signatories of the NPT but also members of the NATO alliance, not only have nuclear missiles stationed on their soil, but also continue to train their troops in the use of nuclear weapons and have operational plans for their use. This is inconsistent with the intent and spirit of Articles I and II of the NPT.

The nuclear-weapon states have not appreciated the stand taken by the non-aligned nations against the discriminatory approach embodied in the NPT. The NPT of 1968 which they continue to promote legitimized the possession of nuclear weapons in the hands of the five nuclear-weapon states, demanding at the same time that others renounce their nuclear option – thus dividing the world permanently into nuclear-weapon and non-nuclear-weapon states. The NPT has also been construed as permitting nuclear-weapon states to disseminate nuclear weapons through their deployment on others" territories, and even to bring pressures on tiny states to accept the stationing of nuclear weapons if the global strategy of a superpower requires it [10]. One also notes the recent pressure on West Germany to accept the modernization of short-range and tactical nuclear weapons as a part of NATO strategy, despite the mounting opposition to them in West Germany [11].

While the nuclear-weapon states and their allies have thus been going ahead with vertical and global horizontal proliferation, and the oceans of the world continue to be infested with submarines and surface vessels carrying nuclear weapons (some of which made special appearances in times of crisis in the Indian Ocean zone and in the South Atlantic zone covered by the Treaty of Tlatelolco), if a littoral country with thousands of miles of coastline acquires, or plans to acquire, a nuclear propelled submarine (without nuclear weapons) there is an out-cry that that is against the spirit of NPT. In the meantime, it has not been possible to convene a Conference on the Indian Ocean as a Zone of Peace, ten years after the relevant Resolution was passed by the UN General Assembly.

Nuclear-weapon states have been advocating nuclear-weapon-free zones (NWFZs) here and there, where it suits their global strategies. They have, however, not been willing to respect their sanctity by agreeing to keep their nuclear-weapon-carrying aircraft and naval vessels out of these zones, or even to declare whether naval vessels carry nuclear weapons when they pass through such a zone or report at a port, or by agreeing not to conduct nuclear-weapon tests in such zones. In this context, one notes that while signing the Protocol II of the Treaty of Tlatelolco, four of the five nuclear-weapon powers have gone on record with their reservations which amount to reserving the option to use nuclear weapons against a member of the NWFZ. The Protocols of the Raratonga Treaty for NWFZ in the South Pacific have been rejected by three of the five nuclear weapon powers, and, further, one of the states in the zone covered by the Rarotonga Treaty, New Zealand, has had to face economic sanctions for deciding not to provide port facilities for ships which do not declare that they do not carry nuclear weapons. These facts cannot be lost sight of by other Third World countries which are being pressurized to form other "nuclear-weapon-free zones". The attitude taken by the nuclear weapons powers towards the rest has been well described by B. Saunders as a taboo: "What you are doing is bad. I do it, and cannot stop doing it. But I shall do what I can to stop you."

5. Threat Perceptions and Confidence-Building Between the North and the South

This attitude cannot but enhance the feeling of insecurity in the non-nuclear-weapon countries, in particular the Third World countries. Given the background of centuries of colonial domination and exploitation, and the current asymmetries in economic and technological strengths, which are sought to be perpetuated through control of international financial institutions including the International Monetary Fund (IMF) and the International Bank for Reconstruction and Development (IBRD) and the inequitable terms of trade including transactions in technology, the Third World countries are led to consider the asymmetry in nuclear capability, which is sought to be perpetuated through the NPT, as yet another lever used by the rich industrialized countries to keep them under continued domination and control. The unilateral actions taken by them in restricting the supply of nuclear technology and materials (for example, the formation in 1977 of the Nuclear Suppliers' Club and the US Non-Proliferation Act 1978), are not only inconsistent with the intent of Articles IV of the NPT and make the present (already discriminatory) NPT regime even more onesided [12], but they also enhance the insecurity of the Third World in the critical area of energy.

Environmentalists have drawn attention to the increasing menace of the greenhouse effect, the need to lower the emissions of CO_2 into the atmosphere, and in this context the need to have a fresh look at the nuclear power option [13]. Nuclear power, through improved engineering designs to enhance safety, better management of radioactive wastes, and better institutional arrangements to prevent diversion of fissile material to weapons, may be expected to have an important role to play in keeping CO_2 emissions down in a world which requires substantially increased sources of power, especially in the Third World. Such a large-scale expansion of nuclear power would, however, be difficult to achieve under the existing regime, with the international community divided into antagonistic blocs, with nuclear weapons deemed legitimate instruments of international power, and a discriminatory regime of technology transactions [14]. It is customary to recommend in the course of NPT discussions, that there should be no discrimination against any NPT members by suppliers in the provision of "exportable" nuclear technology, equipment and material, and that "an equitable code of conduct" for the transfer of technology should be agreed between suppliers and recipients. The existing embargos on commercial transactions in the non-nuclear area (even equipment needed for fundamental research in areas like cosmic rays or high-temperature superconductivity being subject to such embargos), and the stalled negotiations on the Code of Conduct for Technology Transfer at the United Nations Conference on Trade and Development (UNCTAD), do not inspire confidence in such pious recommendations.

Another disturbing trend noticed recently is that of questioning whether the obligations of the nuclear-weapon states under Article VI of NPT were a *quid pro quo* for the obligations of non-nuclear-weapon states under Articles II and III, and of asserting that the obligations under Articles I, II and III (relating to horizontal proliferation in a narrow sense) are far more important than those under Article VI (relating to vertical proliferation). One has only to refer to the Statements of President Lyndon Johnson, Prime Minister Harold Wilson, Premier Alexei Kosygin and others at the time of the negotiation of the NPT, to recognize the disingenuity of such arguments. In any case, such arguments cannot but undermine the confidence of non-signatories and other non-aligned states in the intentions of the advocates of the NPT.

It is clear that the present NPT regime does not and cannot inspire confidence within the non-nuclear-weapon states, especially the Third World countries. While individually several non-aligned countries have agreed to join the NPT because of the coercive pressures from the nuclear-weapon countries, collectively they have refused to endorse it because of its discriminatory character. In the meantime, the coercive diplomacy of the big powers, which are also the nuclear-weapon powers, and their interference and intervention in the affairs of others, continues. The concepts of confidence-building and common security, so assiduously built up in the East-West context, need to be extended on a global basis, in particular to North-South relations. New security doctrines have to be evolved which look at common security on a truly global basis, and do not merely safeguard the interests of the big powers, or of the North.

The First and Second World have yet to adjust themselves to the coming into existence of the Third World as a result of anti-colonial struggles leading to the emergence of over a hundred newly independent nations with their own identities and perceptions of global problems. These nations have their legitimate aspirations to make up for the centuries of foreign rule, and wish to safeguard and consolidate their national independence so as to be able to embark upon a path of rapid development. They are dismayed when the doctrine of nuclear deterrence is given ethnocentric credit for forty "war-free" years when 17 million lives have been lost in over 150 wars since 1945, fought in their countries. The fact that half a dozen "regional" conflicts of long standing have recently started moving towards a resolution as soon as Presidents Mikhail Gorbachev and Ronald Reagan arrived at a certain understanding, shows clearly, if indeed a proof was needed, that these "regional" conflicts were deeply embedded in the conflicts of the big powers in their quest for hegemony. The developing countries were equally dismayed when some of the big powers chose to keep away from the recent UN Conference on Disarmament and Development.

6. Need for New Thinking

In his Address to the 38th Pugwash Conference at Dagomys, Professor Victor Weisskopf [15] referred to the spirit of distrust which still dominated the negotiations:

> "*Ever-increasing deterrence is still the only way to "stabilize" world relations, a situation which will be considered in later times – if there will be people to consider it – as rampant collective mental disease.*"

One must confront this mental disease at its root, rather than attempting to treat symptoms like proliferation with psychological palliatives like the NPT. In the same Address, Weisskopf drew attention to three imminent global dangers: the outbreak of a world war which would soon turn nuclear; the destruction of our environment; and thirdly,

> "*...upheavals in the underdeveloped parts of the world, where both power complexes are trying to expand their influence. Instead of helping these nations effectively to improve their living standards, most of the help is used to acquire weapons against competing groups ...The third danger may perhaps be the hardest to deal with. It is more than the others a political problem. The transition from competition for influence to a collaboration for the improvement of the standard of living, requires a deep change of attitude, perhaps deeper than the change that may lead to an effective dealing with the other problems.*" (emphasis added)

This is where new thinking is needed – more with regard to North-South relations than East-West relations, where there has already been a breakthrough. Above all, we need an integrated approach to disarmament, development and environment.

7. New Pugwash Initiatives: Fourth NPT Review Conference

The need for such an integrated approach was recognized by Pugwash at Dagomys in September 1988 when it issued a special declaration entitled Ensuring the Survival of Civilization [16]. Dagomys also witnessed a beginning of some new thinking by Pugwash on the question of nuclear deterrence. Professor Joseph Rotblat had recently argued that the permanent retention of nuclear weapons, even at a low level, is not acceptable for moral, political and strategic reasons, and had gone on to analyse ways to approach the goal of a nuclear-weapon-free world [17]. The Statement of the Pugwash Council, issued at the end of the Dagomys Conference, explicitly stated a nuclear-weapon-free world was a goal of Pugwash and also emphasized that:

"*...even under an arms control regime that included deep cuts in strategic forces and CTB, the maintenance of any nuclear arsenal whatever, and of doctrines of nuclear deterrence, would pose dangers of nuclear weapons use as well as logical inconsistencies for non-proliferation policy.*"

The Pugwash Statement on the Fourth NPT Review Conference, issued from London on 30 November 1988, goes further [18]:

"*Probably the most unfavourable factor of all is the continued adherence of nuclear-weapon states to explicit or implicit doctrines that insist on the need for maintaining national nuclear forces as deterrents against nuclear or even conventional attack. It needs to be recognized that one of the gravest liabilities of nuclear-deterrence strategy, overt or covert, is its fundamental incompatibility with non-proliferation goals, in as much as there is no logical basis for denying the "right" to a nuclear deterrent to some States while according it to others.*" (emphasis added)

The Pugwash Statement had a number of other aspects:

- it recognized the need to move ahead towards a more robust and more equitable non-proliferation regime, with decisive new actions;
- it expressed concern about the stability of the non-proliferation regime, *inter alia* from continuing vertical proliferation in the form of development and deployment of new nuclear weapon systems by the major powers, and recommended the adoption of *a comprehensive action plan with a specific timetable for stopping and reversing the nuclear arms race* among the major nuclear powers and incorporating, *inter alia,*

  an early agreement on a prescription for including the other nuclear-weapon states in the build-down of nuclear arsenals;

  an immediate moratorium on all nuclear testing, followed by negotiations of a CTB Treaty;

  a universal production cut-off on weapons-grade fissile material under non-discriminatory international control; and rapid progress towards bringing under control the other dimensions of military completion and confrontation.

- it called for a *formal commitment by all nuclear weapon states not to be the first to use nuclear weapons* and not to structure their forces in ways that confer the appearance of first-use intention or capability, and not to threaten the use of nuclear weapons;
- it called for *steps to strengthen the NPT, in preparation for its extension in 1995*, beginning with an immediate initiation of a process, involving non-signatory states as well as signatories, aimed at agreement on means for enhancing the stability and equity of the non-proliferation regime.

What the recent initiatives of Pugwash imply is the recognition that one needs *a fundamental reappraisal* of the approach followed so far with regard to the problems of the nuclear arms race and disarmament. The NPT Review Conference of 1990 will fail in its purpose if it does not face up to the questions discussed above and the recommendations of Pugwash.

In fact, one may go further and suggest a few more items explicitly for inclusion in a time-bound action-plan, so as to build up confidence among the non-nuclear-weapon states:

- an agreement by the superpowers to place the explosive nuclear material (enriched uranium, plutonium and tritium), released as a result of the INF Treaty and (eventual) agreement on deep cuts, in the custody of the IAEA;
- an agreement by nuclear-weapon-states, on a time-bound programme, to withdraw their nuclear weapons and C^3I infrastructures from the territories of non-nuclear-weapon states, as one concrete step towards the gradual liquidation of stockpiles and denuclearization;
- an agreement by nuclear-weapon states to withdraw their nuclear weapons from aircraft, submarines and other naval vessels, and to allow on-board challenge inspections with regard to nuclear weapons;
- an agreement to keep outer space free of all weapons, and a moratorium on all activities of developing, testing and deploying space weapon systems.

8. The NPT Review Conference

William Epstein has reminded us of the main objective of the NPT Review Conferences [19]: "In order to make sure that the nuclear powers were living up to their obligations, the treaty provides for a review conference at intervals of five years". In short, the NPT is not an end in itself, but an instrument that was meant to lead to a cessation of the nuclear arms race at an early date and to nuclear disarmament and a treaty on general and complete disarmament under strict and effective international control (Article VI of the NPT). The objective was and has to continue to be a world from which national arsenals of nuclear weapons and all other major weapons of mass destruction, including the means of their delivery, have been eliminated. This objective, embodied in the very first resolution of the United Nations, unanimously adopted by the General Assembly in January 1946 (and repeated in the Preamble to the NPT), has not only remained unfulfilled, but has not even been taken up for serious negotiations by the nuclear-weapon powers. This objective must be constantly kept in view as the focal point of all discussions of the NPT.

The preparations for the Fourth NPT Review Conference should not, therefore, narrowly confine themselves to the NPT regime as it exists today. *They must be broadened in scope so that the 1990 NPT Review Conference starts a process of reviewing the NPT itself,* and sets in motion a series of negotiations which could lead to the inclusion of the many important elements in the Pugwash Statement[18] so that a strengthened non-proliferation regime may emerge by the time of the 1995 Conference. Such a strengthened non-proliferation regime would have to be a comprehensive system of international peace and security, in which all countries participate on an equal footing. And it would have to be centred on the goal of achieving a nuclear-weapon-free world following a comprehensive action plan with a specific timetable, as emphasized in the Pugwash Statement. Further, it will have to include a multilateral mechanism for monitoring the implementation of the various stages of implementation. All states, signatories and non-signatories of NPT, will have to be a part of this process. One could draw upon the experience of the Law of the Sea Conference to devise an appropriate methodology.

One is reminded that the Pugwash Statement on the Third NPT Review Conference, issued in 1985, had recommended that the Third NPT Review Conference should begin to concern itself with the post-1995 period, and "set in motion the preparation of a new formulation of NPT for discussion at the Review Conference in 1990". Unfortunately, there has been no movement in this direction.

9. Beyond Narrow Non-Proliferation to a World Order Without Nuclear Weapons and War

It has become increasingly clear that the world has to move from the inherently discriminatory doctrine of non-proliferation, which seeks to perpetuate the present unjust and unstable world order, to the objective of a nuclear-weapon-free world; and to do so against a well-adjusted time-frame.

The Russell-Einstein Manifesto has been the credo of Pugwash, but the objective of abolition of war which was projected in the Manifesto has been lost sight of even by Pugwash. It would be useful at the present juncture to remind ourselves of the problem posed in the Russell-Einstein Manifesto:

> *"Here, then is the problem which we present to you, stark and dreadful and inescapable. Shall we put an end to the human race; or shall mankind renounce war? People will not face this alternative because it is so difficult to abolish war. The abolition of war will demand distasteful limitations of national sovereignty ... ."*

The world community has to face up to such limitations of national sovereignty, if peace is to be stable, durable and global. (Actually, it is the big powers that really have to gear up for the "distasteful limitations on national sovereignty" referred to in the Russell-Einstein Manifesto, as for other countries such limitations are part of life). One sees beginnings of movements in this direction in the six-nation initiative and in the Delhi Declaration on Principles for a Nuclear Weapon Free and Non-Violent World, signed by Prime Minister Rajiv Gandhi and General Secretary Mikhail Gorbachev on 27 November 1986 [20]. These have to be pursued vigorously.

While discussing the difficulties involved in securing the complete elimination of nuclear weapons, Rotblat dealt with the concern as to how to ensure that one side will not cheat in the last stages of nuclear disarmament, and conceal a small number of such weapons with which it could gain political power, if not world domination. In this context, he suggests the setting up of an international authority endowed with a small nuclear weapons capability [17]:

> *"The United Nations has proved itself in a variety of policing duties and peace negotiations, and has acquired enough status to be entrusted with an even mightier task, to be the custodian of an instrument designated to prevent a threat to world security ... the treaty abolishing national nuclear arsenals should be preceded by the setting up of [this] international depository of nuclear weapons and means of delivery, sufficient to deter a potential violator of the treaty."*

The world will need other new international instruments. A world without nuclear weapons will not be more safe than the world in 1939, unless we have adequate instruments for conflict resolution and peace-keeping. The United Nations will, therefore, have to be suitably strengthened so as to be able to meet any threat of aggression with a standing international peace force. The Parliamentarians Global Action Group has made several other recommendations which deserve serious consideration [21]. For example, they propose an expanded system of world courts and institutions for mediation and arbitration, as an alternative to the battlefield as a way of setting disputes; and a global multilateral verification system for monitoring the gradual demilitarization of nations. As it has pointed out, while more work needs to be done on the detailed problems of arms reduction, peacekeeping and enforcement, verification and conversion to a peacetime economy, the key elements were already outlined in a series of governmental and intergovernmental documents produced since the Second World War, including the principles of a comprehensive disarmament treaty agreed to by the United States and the Soviet Union in 1961 (the McCloy-Zorin Principles), the USA and the Soviet draft disarmament treaties of 1962, and the Final Draft Document of the UN Special Session on Disarmament in 1978. These could be the starting point for fresh negotiations, involving all nations. And such negotiations will have to start soon, in parallel with the negotiations on the comprehensive action plan with a specific timetable for stopping and reversing the nuclear arms race.

Without such measures, considered conjointly with serious efforts towards a just world order, any peace or disarmament agreement is bound to be fragile. Durable peace and global common security demand an equitable and just world order. The fight for such a world order has to find a place on the agenda of any genuine efforts towards disarmament, peace and global security.

References

1 For a brief history of this early period, see B. Sanders, "The Treaty on the Non-Proliferation of Nuclear Weapons and the Relations between Superpowers", in *Nuclear Non-Proliferation*, M. P. Fry, N. P. Keatinge and J. Rotblat (eds), Springer, Heidelberg, 1990

2 For details see "The Non-Aligned Movement and Disarmament", in *Surviving the Second Millennium: Disarmament and Information: a Research Monograph for UNESCO by Namedia Foundation*, (New Delhi, 1987). For a brief summary, see B. M. Udgaonkar, "Non-Aligned Movement and Pugwash", *New Times*, (Moscow) August 1988, special supplement.

3 Resolution 2028(XX) adopted by the General Assembly of the United Nations on 19 November 1965

4 Statement of the Pugwash Council in *Global Problems and Common Security*, J. Rotblat and V. I. Goldanskii (eds), Springer, Heidelberg, 1989, p. 255

5 5.W. M. Arkin and R. Fieldhouse, *The Nuclear Battlefields: Global Links in the Arms Race*, Ballinger, 1985

6 P. Bracken, *The Command and Control of Nuclear Forces*, (New Haven, Connecticut, 1983)

7 J. Singh, "Nuclear Arms Race: Dimensions and Trends", paper presented to the International Conference "Towards a Nuclear Weapon Free and Non-Violent World", New Delhi, 14–16 November 1988

8 See L. Spector "New Players in the Nuclear Game", *Bulletin of the Atomic Scientists*, January–February 1989, for a recent review of this aspect. See also R. Bolt, "Plutonium for All: Leaks in Global Safeguards", *Bulletin of the Atomic Scientists*, December 1988

9 D. A. Fischer, "Drawing the Threshold States into a Regime of Restraint, by Joining NPT or Otherwise", in *Nuclear Non-Proliferation* see ref. [1]

10 For the problem faced by the tiny coral reef island of Palau in South Central Pacific (population 15 000), see Jennifer Scarlott, *Bulletin of the Atomic Scientist*, November 1988

11 S. Adam, "Nuclear Missiles: Bonn's Allergy", *Times of India* (Bombay) 26 April 1989

12 V. Knapp, "Non-Proliferation Treaty can and should be strengthened", *Proceedings of the 38th Pugwash Conference*, Dagomys, September 1988, pp. 248–50

13 See, for example, Report of the Energy Working Group at the International Conference on "The Changing Atmosphere: Implications for Global Security", Toronto, 27–30 June 1988

14 K. Subrahmanyam, "Alternative Security Doctrines", paper presented to the International Conference "Towards a Nuclear Weapon Free and Non-Violent World", New Delhi, 14–16 November 1988

15 V. F. Weisskopf, Address to the 38th Pugwash Conference, Dagomys, 29 August 1988, *Pugwash Newsletter*, October 1988

16 "Ensuring the Survival of Civilization", in *Global Problems and Common Security*, see reference [4] p. 259

17 J. Rotblat, "The Elimination of Nuclear Weapons: Is it Desirable? Is it Possible?" (Chapter in the Pugwash Verification Book, to be published)

18 Pugwash Statement on the Fourth NPT Review Conference, see Appendix B, this volume

19 W. Epstein, "The Proliferation of Nuclear Weapons and Efforts to Prevent Further Proliferation", *Proceedings of the 37th Pugwash Conference*, Gmunden am Traunsee, September 1987

20 Delhi Declaration on Principles for a Nuclear Weapon Free and Non-Violent World (27 November 1986)

21 Parliamentarians Global Action, "Global Security", paper presented to the International Conference "Towards a Nuclear Weapon Free and Non-Violent World", New Delhi, 14–16 November 1988

The Linkage Between Non-Proliferation, Deterrence Policy, Nuclear Testing and the Arms Race

Lawrence Scheinman

1. Non-Proliferation and the Comprehensive Test Ban

A major focal point for strengthening the Non-Proliferation Treaty (NPT), which is the central legal and political basis of the non-proliferation regime has been negotiation of a comprehensive test ban (CTB). Achievement of a CTB would accomplish a number of things.

First, it meets, more directly than virtually any other arms control measure, the concern expressed at all three Non-Proliferation Treaty Review Conferences about the inadequacy of nuclear-weapon state efforts to meet their political obligations under Article VI of the NPT. Secondly, and related, insofar as non-proliferation is concerned, it would have political significance beyond any military significance, because it would apply to the three nuclear-weapon state parties to the NPT (the United States, the Soviet Union and Great Britain) as well as to non-nuclear weapon state parties, and thereby help to overcome the concern with discrimination that has been raised against the NPT by a number of countries. Parenthetically, it would deny states which have argued that NPT perpetuates an unacceptable *status quo* a valuable piece of evidence to support that claim. Thirdly, beyond establishing a legal barrier, it would establish an important psychological barrier to testing by a non-nuclear-weapon state whether or not party to a CTB treaty, and could not help but have some impact on France and China, the two nuclear-weapon states that have not even joined the Partial Test Ban Treaty. In this conjunction, existence of a CTB would reinforce the hand of those in non-NPT states who oppose acquisition of nuclear weapons.

These are three powerful non-proliferation reasons favouring agreement on a comprehensive test ban. But even if the United States and Great Britain were to agree that the above considerations are valid ones, they are not, at this time, prepared to agree to a CTB on the ground, correct or not, that testing is necessary to maintain and to modernize their nuclear weapons (deterrence) systems. The Final Document of the 1985 NPT Review Conference, while calling for trilateral negotiations to begin in 1985 to achieve a CTB, noted that:

> "*... certain states party to the treaty, while committed to the goal of an effectively verifiable comprehensive nuclear test ban treaty, considered deep and verifiable reductions in existing arsenals of nuclear weapons as the highest priority in the process of pursuing the objective of Article VI.*"

This language was inserted at the request of the United States and Great Britain.

While it is clear that negotiations of a CTB would be a major reinforcement of non-proliferation, there is no assurance that nuclear proliferation would be ended. Some states argue for a CTB knowing that the probability of it being agreed in the near term is very low, but that they can deflect attention from their own rejection of NPT by focusing on the deficiencies of the NPT. In arguing for a CTB they can appear to be genuinely concerned about the risk of proliferation, can embarrass the superpowers for failing to "meet their obligations", and can contend that they will remain outside the NPT unless and until its discriminatory features are eliminated. Continuing nuclear weapon tests is only one aspect of discrimination, however. The presence of large nuclear arsenals in the hands of the nuclear-weapon states is another, and the existence of nuclear weapons could easily become another reason for holding out from NPT participation even in the face of achievement of a CTB.

Even more important in terms of the impact on proliferation of a CTB is the fact that the approach a state takes towards the question of whether or not to forswear nuclear weapons, and to legally obligate itself to non-proliferation in a multilateral treaty, depends on complex security and political considerations. The existence or non-existence of a CTB may be only marginally relevant to how a state evaluates the policy it should choose. For a state to endorse and adopt non-proliferation it must conclude that its national security and political interests are best or better served by joining in non-proliferation than by maintaining a nuclear option or even acquiring nuclear weapons.

This judgement is a function of several considerations: the regional security situation of a country; its alliance relationships; the character of the domestic political debate on the issue; the importance to it of countries about whom it is greatly concerned to maintain good relations; and the extent to which the latter make friendship, support and diplomatic and economic interaction dependent on the nuclear policy of the state in question.

In the final analysis, many of these judgements may depend on how nuclear weapons are perceived in terms of their military utility, their political value and their legitimacy. This confronts us with the question whether nuclear testing, modernization, and maintenance of weapons arsenals reflect the view that nuclear-weapon states perceive their status as having military and/or political value for promoting national interests. It also raises the question of the linkage between nuclear proliferation and deterrence policy.

2. Deterrence Policy and Non-Proliferation

The question has been raised why, if deterrence was considered good and useful for some states, it should not be for others? This is an important and complex issue. Several points deserve emphasis:

First, it is not so clear that deterrence is all that good, useful and valid. For deterrence to be considered successful, a challenge must be averted in the first place. This depends on the capability and the will of the would-be deterrer. He must be seen by the potential adversary as capable of inflicting damage that makes the costs of challenge greater than any possible gain; and he must be seen as having the political resolve to inflict that damage if militarily challenged – that is if he is challenged even after making a well-defined and well publicized commitment. Several studies of deterrence show that deterrence does not always work[1]. The Cuban missile crisis is often cited as a case in point – despite clear and unambiguous warnings that the United States would not tolerate missiles in Cuba, the Soviet Union deployed them, bringing the superpowers closer to the brink of war than ever before or since. The challenge was made at a time of undoubted US military superiority. The 1973 attack by Egypt on Israel is also cited as an example of deterrence failure. Although not involving nuclear deterrence that case is cited as demonstrating how intangible factors, like national honour, emotion, anger and so on, can drive decisions that would not have been taken had "reason" prevailed. Deterrence presumes rational actors, but decisions of seemingly rational people are not always made on so-called "rational grounds". So deterrence may not lead to the outcome it is intended to achieve. In short, situations may make intentions or even capabilities irrelevant, making rational choice meaningless. This is, generally speaking, a good reason for all to reduce reliance on nuclear weapons.

Secondly, even if it is considered to be largely true that deterrence, that is nuclear deterrence against nuclear attack, has worked between the superpowers, there is no *a priori* reason that it will work elsewhere. That is to say, it is very unclear that deterrence is *transferable*. There are several attributes of deterrence between the United States and the Soviet Union whose existence should be recognized and not underestimated:

- The two countries do not share common borders along which the opportunity for direct contact and possibly conflict can arise.
- The two countries do not have deep historic animosities and have never fought a war with one another.
- There are no revanchist claims of one against the other.
- Both countries have invulnerable second-strike capabilities which assures that whoever strikes first will suffer incalculable damage through retaliation. In recent times the Soviet Union, reversing an earlier approach, has come to conclude that it is desirable and even in the Soviet interest that the United States have a secure second-strike retaliatory force. This, along with

acceptance of "sufficiency" rather than "superiority" or even "parity" has significant implications for stability.

One could add to these points the command and control systems of the nuclear forces and the stability of government, although some might quite properly question whether the latter point can be sustained as a unique or different quality.

The four listed factors raise serious questions about how transferable the working deterrence system is to any other situation. With respect to the fourth point, in particular, how plausible is it that a country would be allowed by neighbours to achieve an invulnerable second strike capability before being preempted by those neighbours? The Israeli early strike against an Iraqi reactor that was not even yet operating when it was destroyed underscores the risks of pre-emption and the improbability of successful transference of deterrence.

These points notwithstanding, it is difficult to arrive at unambiguous conclusions on the question of linkage between deterrence policy and non-proliferation. There is little doubt that the proliferation effect of deterrence policy can work differently for different countries. First, for those states which are protected by what they perceive to be credible security guarantees by a nuclear ally, deterrence is a reason *not* to consider proliferation; as long as states feel politically comfortable with the credibility of the nuclear guarantee, security can be purchased at lower cost and with lower risk through nuclear alliance than through nuclear proliferation. A breakdown in current security arrangements or an alteration in perceptions about the credibility of extended nuclear deterrence could lead to reassessment on the part of protected countries regarding nuclear weapons or a nuclear option. Secondly, for unprotected states, or states which consider nuclear guarantees by a nuclear-weapon state non-credible or inadequate, deterrence policy can glitter like a diamond in the rough and if seen to be effective in precluding military challenge (or at least successful military challenge) may be an *incentive* to acquire nuclear weapons. Foreign policy behaviour of nuclear-weapon states involving overt or implied nuclear threats, or the demonstrated utility in a conflict of the military use of nuclear weapons, would only reinforce the incentive, even if, for reasons discussed above, it is questionable whether the acquisition of nuclear weapons would enhance national security.

To summarize, there are reasons to question whether deterrence really works in the first instance, and to doubt that nuclear deterrence, if it does work in the superpower case, would work as well in other dyadic cases, and therefore to question the wisdom of considering adopting a nuclear deterrent approach. Nevertheless, one cannot deny that nuclear weapons appear to have contributed to prudence in US-Soviet relations and that war between the two powers has been avoided. The central point is to read the right lessons from this. While some would argue that the lesson is the validity of nuclear deterrence, a stronger case can be made that the attributes of the US-Soviet relationship described above are very critical to the situation, and that it is not the presence of nuclear weapons that has created a stable balance, but the assured second-strike capability which has enhanced deterrence. A small, vulnerable nuclear force could work quite differently, decreasing security and inviting or encouraging pre-emption.

Since lessons of deterrence may be misread or misinterpreted, but the continued presence of deterrence policies and of the nuclear arsenals and testing programme may encourage non-nuclear weapon states to continue, or to adopt policies of maintaining a weapons option, there is substantial reason to seek to reach beyond deterrence and to replace it with something more stabilizing and more durable, such as policies of reassurance. Subject to the caveat that *on ne detruit que ce qu'on remplace* that should be at the top of the international security agenda of the 1990s.

References

1 R. N. Lebow and J. G. Stein, "Beyond Deterrence", *Journal of Social Issues*, vol. 43, no. 4 (1978)

Part V

The Arms Trade and Regional Conflict

Recent Patterns of Arms Trade and Regional Conflict

Saadet Deger

1. Introduction

During the last four decades almost all regional conflicts which have erupted into wars have been waged among the poorer countries of the world. These so called "Third World wars" have increasingly been fought with sophisticated weapons which are either imported or produced domestically. However, even in the latter case, the technology is invariably imported. Arms imports have acted as a catalyst to destruction as incipient conflicts have erupted into explicit hostilities. The principal purpose of this paper is to analyse the trends in arms sales to the Third World and relate them to the course of conflicts.

We concentrate on developing countries primarily because they have been subject to the actual wars that have been fought round the world. In addition, these countries, taken in aggregate, dominate the global purchase of arms. On the other hand, the overwhelming proportion of sales emanate from the developed countries, both East and West. There is, therefore, a close North-South dimension to the problem intricately related to the East-West nexus. The discussion also shows that political, economic and military factors are interrelated and that often the causal connections are not easy to unravel.

2. Quantitative Trends of the Arms Trade

The share of the Third World (defined here as developing countries outside the main Alliances: the North Atlantic Treaty Organization and the Warsaw Pact but excluding China) in total value of arms trade *for major conventional weapons*, has been declining over the last few years though not at a rapid pace. In 1983 the Third World purchased 69.2 per cent of the global total arms sales (major weapons); by 1988 this ratio had fallen to 61.4 per cent. The trend has only partly to do with suppliers' restrictions, since arms trade controls have not succeeded in achieving their stated objectives, except in a few well-publicized embargoes. Even for the latter cases, there have been significant evasions: witness the Iran-Contra affair, or even the most extreme case of South Africa, where the "history of implementation" has become a "history of disimplementation" of the embargo. At the same time, the number of suppliers have increased and there are now more competitive pressures which serve the interests of the buyers and under usual circumstances should help in increasing demand. This has not happened. The main reason for the decline in Third World share is the lack of funding, either due to the non-availability of foreign assistance and loans (net of debt service), or due to insufficient foreign exchange during a period of global crisis.

It should be noted, however, that the foregoing decline occurred only for major weapons systems, which are the only ones reflected in the data base of the Stockholm International Peace Research Institute (SIPRI). When data from the US Arms Control and Disarmament Agency (ACDA) are taken into account the share of the Third World in all arms imports has remained stable at around 80 per cent from 1980 to 1986 (the latest year for which data are available) [1]. This again emphasizes the role of financial constraints. Major weapons systems tend to have much higher unit and aggregate costs; hence their demand falls faster during periods of austerity. An additional reason may be that high-intensity conflicts are more dependent on major weapons. In spite of the Iran-Iraq war, the frequency of such conflicts may have been declining in the Third World. The number of major armed conflicts [2] at the beginning of 1988 was 33; by the end of the year this had declined to 28.

Since our main interest is to delineate important trends and relate them to regional conflicts, which tend to be high-intensity ones using sophisticated weapons, we will concentrate on the trade in major weapons. Extensive use is made of the SIPRI data base which is designed to give a quantitative index of the aggregate volume of arms imported by the Third World as well as specific regional breakdowns. Henceforth all data and figures relate to the trade and transfer of major conventional weapons as well as licensed domestic production (to incorporate technology transfer).

Table 1. Growth Rates of Arms Imports, Annual Average Rates (per cent)

Regions	1970-79	1980-88
Middle East	3.4	1.2*
Iran-Iraq	12.0	8.6
(for 1970-78)	(24.1)	n.a.
South Asia	5.0	16.2
Far East	1.9*	7.0
North Africa	41.1	-27.3
Sub-Saharan Africa	20.2	0.5*
Central America	6.9	-5.0
South America	16.7	-15.6
Third World	8.6	0.7*
Third World (without Iran-Iraq)	7.6	-0.5*

Source: For method of calculation see Appendix. In particular, note that numbers with asterisks (*) should be considered as zero growth from a statistical significance point of view. For the names of countries in each region see Stockholm International Peace Research Institute, World Armaments and Disarmament: SIPRI Yearbook, 1989 (Oxford, 1989), Appendix 6A.

A predominant feature of the world arms market, in the last two decades, has been one of rapid growth followed by stagnation. The 1970s showed arms imports by developing countries increasing at a rapid rate. The annual growth rate of Third World arms importation between 1970 and 1979 was 8.6 per cent per annum. During the current decade, in the period 1980 to 1988, the growth rate has been almost zero. If the Iran-Iraq war related arms imports are removed the trends reveal a decline; the growth rate for 1980–1988 was probably negative (though in terms of statistical significance it was not different from zero). The precise figures are summarized in Table 1.

The aggregate trend holds through for most of the regional variations, even though there are important exceptions. The most astonishing declines in the decadal growth rates are to be seen for North Africa, Central and South America and Sub-Saharan Africa (see Table 1). For these four regions, comparing the 1970s and the 1980s, the change in the growth of arms imports show significant structural variations: for North Africa a 41 per cent per annum increase was translated into a 27 per cent per annum *decrease* as between these two periods; Central America saw a positive growth rate of 7 per cent transformed to a *negative* growth rate of about 5 per cent; South America, which had witnessed a 17 per cent annual increase during the 1970s now reversed the trend totally with arms imports *declining* by 16 per cent per year; and finally, Sub-Saharan Africa reversed an annual increase of around 20 per cent to a stable flow of *zero growth.* Even Iran and Iraq who fought a major protracted war between 1980 and 1988 had higher arms import growth during the previous decade as compared to the war period. The two important exceptions were South Asia and the Far East. Long-standing regional military rivalries, occasional tension and open conflicts, arms races, superpower involvement and relatively stable economic conditions, all contributed to large-scale Asian arms purchases which kept the market buoyant.

The level of Third World arms imports, as opposed to its growth, has stabilized at a relatively high level for the last ten years. Around 1977 the value of arms imported (in constant prices) by developing countries increased by around 50 per cent from the previous year's level; in 1985 prices this amounted to over $22 billion. It has remained consistently at a similar level with only one annual exception, that of 1987. If the exceptional importation forced by the Iran-Iraq war is excluded, the total value would be around $20 billion in 1985 prices. Using US Department of Defense procurement price deflators [3], this $20 billion can be given a very approximate value of about $23 billion at *current* (1989) prices.

The total amount does not seem to be very high, in terms of economic costs, even though the costs in terms of social and human destruction is incalculable. The Third World's total import of all merchandise goods in 1985 was of the order of $360.5 billion [4] (the figure excludes China since it is not included in the SIPRI Register as an importer). Arms imports of $20 billion then accounts for less than 6 per cent of aggregate imports. By any criterion this is not extravagant.

Yet, the alternative or opportunity costs of such imports, even if they are not directly paid for, can be very high indeed. One way of looking at the problem is to compare it with the imports of machinery and transport equipment. This category of foreign imports contribute the most to the Third World economic growth and very often acts as a binding constraint on further development, since domestically-produced substitutes are not generally available. On the other hand, weapons are most often in competition with machinery and fall within similar industrial trade classification. In 1985 it can be estimated that Third World machinery imports were of the order of $108 billion [4]. Therefore, the purchase of foreign major weapons was equivalent to around 20 per cent of all machine imports. In a counterfactual scenario, if arms imports were diverted to machine imports, the latter would rise by 20 per cent. The effect on economic growth of such a transfer would be exceedingly beneficial for developing nations.

Table 2. Regional Allocation of Arms Imports by Third World Countries (US $ billion at constant 1985 prices)

	1970-74	1975-79	1980-84	1985-88
Middle East	32.8	37.2	51.7	42.0
South Asia	5.7	8.5	12.0	18.3
Far East	14.5	14.1	12.8	14.9
North Africa	1.7	16.7	11.9	3.4
Sub-Saharan Africa	4.1	8.9	8.4	7.3
Central America	1.2	1.6	3.4	1.8
South America	5.8	10.3	13.6	4.4

Source: SIPRI data base.

The regional allocations of arms imports are shown in Table 2. A general feature is observed. For almost all regions there was a bunching of imports between 1975 to 1984, and then a considerable fall or stability in the volume of transfers. In addition to explaining the earlier discussion on decadal growth rates, another important characteristic becomes clear. Arms imports tend to rise during a period of economic expansion, as was the case for the Third World during 1975–1985. It should also be emphasized, to put the arms trade in perspective, that it is highly concentrated on the demand side. Only 15 countries, out of a total of about 66 for which arms transactions are recorded in the SIPRI data base, constitute 75 per cent of total Third World demand. During the period 1984–1988, the top five in the importation league accounted for around 50 per cent of the global purchase of armaments to the Third World. Table 3 gives the demand shares of the 15 largest importers. For the remaining 50 countries or so, listed as others, there has been a decline in import volume by around 50 per cent between 1984 and 1988.

Next we discuss the trends as regards the suppliers. The two largest exporters are the Soviet Union and the United States with France, China, the UK and West Germany making up together the largest six which overwhelmingly dominate the global total exports to the Third World. Several developing countries are active participants in arms exports, particularly Brazil, Israel and North Korea. Their market share is low but by introducing more competitive tendencies these countries make it easier for demanders to get some of what they want even in the face of export regulations.

Table 3. Share of Third World Imports by Countries 1984-88 (per cent)

Countries	Shares
Iraq	13.9
India	12.4
Saudi Arabia	8.1
Egypt	6.9
Syria	6.2
North Korea	5.0
Angola	3.8
Pakistan	2.9
Iran	2.9
Libya	2.7
Taiwan	2.7
Israel	2.5
South Korea	2.0
Afghanistan	1.9
Argentina	1.8
Others (more than 50 countries)	24.2

Source: Calculated from SIPRI data base.

Table 4. Value of Arms Exports to the Third World: Six Major Suppliers (US $ billion at constant 1985 prices; bracketed terms are shares in total) 1984-88 (per cent)

	1970-74	1975-79	1980-84	1985-88
Soviet Union	26.9	35.1	38.3	38.4
	(40.8)	(36.1)	(33.7)	(41.8)
United States	23.8	35.8	29.9	18.6
	(36.1)	(36.9)	(26.3)	(20.2)
France	5.1	10.4	15.2	11.4
	(7.7)	(10.7)	(13.4)	(12.4)
China	1.5	1.5	3.9	6.5
	(2.3)	(1.6)	(3.4)	(7.1)
Great Britain	5.3	5.7	5.3	5.4
	(8.0)	(5.8)	(4.7)	(5.9)
West Germany	0.6	1.1	3.7	1.8
	(0.1)	(1.1)	(3.3)	(1.9)
Total	65.8	97.2	113.7	92.0
	(100.0)	(100.0)	(100.0)	(100.0)

Source: SIPRI data base and author's calculations.

Table 4 gives the value of arms exports, for five year periods spanning 1970–1988, for the different groups of exporters mentioned above. During the 1980s the two largest superpowers have a stable market share of around 60 per cent; however, it is lower than in any period since the Second World War. Compared to the 1970s they have seen their relative importance in the export market declining from about 75 per cent to 60 per cent. The next four "majors" have in aggregate seen their market share rising as between the two decades from around 18 per cent to about 25 per cent. This change of relative position between the two largest exporters and the remaining four large sellers has important implications. These are discussed in the next sections.

Table 5. Growth Rates of Arms Exports to the Third World by Major Countries, annual average rate (per cent)

Regions	1970-79	1980-88
Soviet Union	7.5	3.3
United States	5.2	-5.0
France	16.6	-2.9*
Great Britain	2.9*	6.6
West Germany	37.3	-2.5*
China	2.9*	19.2
All other countries	20.7	-1.9

Source: For method of calculation see Appendix. In particular, note that numbers with asterisks (*) could be considered as zero growth from the point of view of statistical significance.

Export patterns clearly follow the structural shifts in the market. Growth rates fluctuate on a year-to-year basis but some consistency is achieved by looking at decadal averages. Table 5 shows the growth of arms exports over 1970–79 and 1980–88 for the six largest exporters. There is a significant decline in the latter period, compared to the former, for almost all countries among the big six. This is primarily due to overall market stagnation, as well as the implicit export controls employed during the Iran-Iraq war. The only exception is China which saw a dramatic upsurge in its arms exports during this decade. The growth of Chinese arms exports went up dramatically from an annual 3 to 20 per cent between the 1970s and the 1980s. It profited hugely from the Iran-Iraq war, supplying to both combatants [5]. It is also a major supplier to Pakistan and its exports have taken advantage of the arms race in the Indian sub-continent over the last ten years. The UK exports have remained quite stable on a long-term basis but shown strong fluctuations in terms of annual values. From about 1971 its exports have hovered around the one billion dollar mark (in constant 1985 prices), but there are significant individual outliers which tend to distort the statistical picture. The growth rates, therefore, are subject to major changes depending on which year(s) one chooses to begin (and end) with. Overall, in spite of the encouragement of the Margaret Thatcher Government towards greater arms sales, the total value of UK exports have been stable during the present decade. Once again this reflects the level of market stagnation rather than the political will to curb the trade in weapons.

3. Qualitative Features of the Arms Trade

Equally important as the quantitative trends are some of the prominent qualitative features of the arms market. The first is, of course, the political control that exporting countries try to impose on transfers. Given the lethal nature of the product, the possible misuse of arms in violent conflict and the degree of uncertainty surrounding recipient intentions, it is but natural that governments of supplier countries will be wary about the nature of transfers. The 1980s have seen increasing awareness of proliferation problems and the necessity of certain forms of institutional restraints. The 1988 United Nations Special Session on Disarmament debated this issue in some detail though it failed to reach a definitive conclusion, nor did it suggest specific policy rules to curb the trade.

The main reason for the failure of arms trade controls is not necessarily because the exporters lack the political will. The real problem lies in the subtle transformation of the nature of the trade itself. Economic and commercial factors are gaining in importance as the number of countries willing and able to supply major weapons systems increase rapidly. With increased competition, the decline in the market shares of the superpowers, less market concentration and greater technology transfer at a price, commercialization threatens the political framework within which the market has traditionally operated. Arms transfers are now a matter of both politics and economics. This phenomenon requires special attention. Hence the next section deals in much more detail with the political economy of the arms trade.

Even though the volume of sales has stabilized, vertical proliferation in the form of more sophisticated weapons systems has added a new dimension to the fundamental problems of the arms transfer process. The spread of nuclear weapons through the trade in technology and raw materials is now well documented. Israel and South Africa are now reputed to have a modest but deadly arsenal; India and Pakistan are reported to have the capability of producing a stockpile in a very short period of time if required. In the same vein the "poor nation's nuclear weapons", that is chemical weapons, are increasingly being acquired in the Third World. Their use during the Iran-Iraq war, as well as for domestic repression, is well known.

Maximum recent concern in the field of proliferation has been expressed by the sale of ballistic missiles, as well as the transfer of production technology which allows conflict-prone countries to manufacture them indigenously. In addition to conventional weapons these missiles can carry chemical and biological warheads as well as enhancing potential nuclear capability. It is now estimated that about 20 countries possess ballistic missiles of various ranges. As the "war of the cities" in the Iran-Iraq war showed, such missiles can create large casualties in regional conflicts among neighbours, particularly for civilians. If such missiles were armed with chemical, biological or nuclear warheads the implications could be horrifying.

The final qualitative feature of note is the rapid increase of arms producers in the Third World fuelled by the trade in technology as well as by the development of industrial capacity among Less Developed Countries (LDCs). In the early 1980s about 15 developing countries had a reasonable volume of arms manufactures within their economies. This itself was a substantial rise from the situation about a decade earlier. By the late 1980s the number has risen to 25 countries which have progressed beyond the manufacture of small arms and have become capable of producing more sophisticated products. The majority of arms manufactured are through licensed production or through co-development. Except for one or two countries, like Brazil or Israel, most LDCs are heavily dependent on technology transfers to set up domestic arms industrialization. The parallel movement, of the trades in the final product as well as the capability to produce that product, is now a well-established feature of the market.

Among the suppliers of technology, the commercial exporters, such as France, the UK or the FRG, have been more forthcoming in helping LDCs to set up their own domestic arms industries. The two superpowers have been more careful about technology proliferation, though the United States has provided some help in these respects. The most circumspect has been the Soviet Union. It has provided only India, in the Third World, with blueprints to manufacture state-of-the-art weapons and utilize Soviet technology to create a powerful domestic base. For all other client states the Soviet Union has only supplied the final product rather than adding to their domestic production capacity.

4. The Political Economy of Arms Transfers

There can be no doubt that the arms trade is a matter of crucial military significance and has a close relationship with national and regional security as well as with the process of establishing peace. The importing countries certainly believe that their demand is motivated by security considerations alone, which in turn determines the welfare function of the state. Economic factors constitute the constraint within which the government has to operate. If the economic environment is favourable (say after the oil price boom), or it is possible to maintain socio-economic expenditure while continuing to receive foreign arms (through military assistance), then inevitably arms purchase rises with perceived security threats.

More controversial is the relative role of political and economic factors that motivate *arms sellers*. Here the position is not at all clear cut as to whether the major part of the trade is motivated by political concern ("to make friends or influence people"), or whether the trade is fuelled by the desire to make commercial gains ("the merchants of death"). Clearly both factors will play a role; the analytical discussion centres around which is more important in a relative sense.

It should also be noted that the issue is relevant at the level of government-to-government transactions only. All firms and industrial organizations that produce and sell the weapons clearly do so to make profits alone. Considerations of domestic national security is a matter for the government alone. Hence, all governments in the post-War era have sought to impose some form of controls on exports, though the black market is again on the rise.

The central importance of political factors was stressed by A. J. Pierre in 1982 [6]: "Arms sales are far more than an economic occurrence, a military relationship, or an arms control challenge – *arms sales are foreign policy writ large*". Arms sales may promote the security of allies and ensure the stability of friendly governments; they may be used by supplier nations to gain influence and to control recipient countries; and they may be utilized to balance an existing arms race and to "stabilize" regional disparity between allies and others. All of these factors, in various combinations, have been used by the sellers to justify their weapons transfers.

But the economic motives are always present, either in explicit form where arms, though sold to allies, still need to be paid for, or in implicit form where arms exports are sanctioned to countries which have a dominant importance in trading and financial relationships. As SIPRI stressed in 1982 [7]:

> *"When an economy is in crisis, more weight is allocated to economic than to political arguments. With high unemployment, foreign trade imbalances and budget deficits, this is now particularly evident in the West European arms manufacturing countries ... arms exports improve balance of payments, lower unit prices through the advantages of scale, and ensure employment in the arms industry."*

A simple though not exact distinction would be that the two superpowers are predominantly interested in political hegemony when they determine their arms transfer policy, whereas the other exporters are mainly concerned with economic and commercial gains, though of course arms will never be sold to "enemies". (Great Britain did sell to Argentina in the early 1980s though the flow stopped after the Falklands/Malvinas war). This distinction can be used as a working hypothesis, though major exceptions will remain. For example, during the late 1970s and early 1980s, the three principal sources of hard currency earnings for the Soviet Union were gold, energy and armaments. The export of weapons to oil-rich states, particularly in North Africa, was a significant boost to Soviet dollar earnings at a time of considerable balance of payments problems [8]. The United States raised its direct sales of arms from $1.1 billion in 1970 to $16 billion in 1981 [9]. The superpowers are not immune to the lure of export earnings if they can find customers willing to pay the price. But in a stagnant market with increased competition and falling shares (as discussed earlier) commercial motives for the two largest supplying powers take on secondary importance. Currently, their main motivation seems to be political even though some economic gains are made [10].

For the other countries the dominant motive for firms and government alike is to make profitable transactions. There are now innumerable ways in which transactions can take place: direct sales through cash; loans which finance the payment involved, with varying types of debt servicing methods; offsets utilizing technology transfer mechanisms; and other innovative forms of Compensatory Trade Agreements (CTAs). A classic example is France. As E. A. Kolodziel wrote in 1983 [11]:

> "*...the transfer of arms and technology is a critical component of an overall economic strategy. This strategy has as its main elements the modernization of the French economy; the preservation of full employment, with particular emphasis on employment in sections of advanced technology; the pursuit of economic growth; the importance of France's international competitive position, so as to avoid balance of payment deficits; the development of a global market for French goods; and adequate access to raw materials, especially oil.*"

The British government vigorously promotes defence exports. In spite of high domestic procurement spending, around 20 per cent of military production is exported and industrial viability depends crucially on overseas sales. Even West Germany, relatively concerned at the security implications of arms transfers in the past, has increasingly become embroiled in commercial motives; as U Albrecht claimed in 1986 [12] *Foreign policy considerations are losing their importance. In a dispute over a potential weapons sale, it is increasingly likely that economic motives will prevail over diplomacy.* It has been claimed that European firms often export 50 per cent of their output while a country like Brazil could sell 70 per cent of domestic production abroad.

Table 6. Percentage Market Shares of Total Arms Exports to the Third World

	1970-74	1975-79	1980-84	1985-88
Predominantly political exporter countries	76.9	73.0	60.0	62.0
Major commercial exporter countries	18.1	19.2	24.8	26.1
Other exporters	5.0	7.8	15.2	11.9
Index of market concentration	1.6	1.74	2.25	2.14

Source: Calculated from Table 3. The first group consists of the superpowers; the second group consists of France, China, UK and the FRG; all other exporters make up the rest. For the index of concentration see Appendix.

Table 6 gives the market share of three groups of exporters: the predominantly political sellers are the two superpowers though some commercial considerations are also present here; the mainly commercial sellers constitute the rest of the exporters; the latter countries can be sub-divided into two groups: the four majors and all the other smaller exporters. This last group also contain some Third World countries themselves (like Brazil and Israel). The superpowers currently enjoy 60 per cent of the market while the commercial sellers have a share of around 40 per cent. However, we also need to consider the fact that part of superpower sales are also for cash (or favourable credit). Hence, the *economic motivation could account for 50 per cent of the total exports of arms*, that is half of the total market evolves around profitability and allied considerations. This is a sizeable figure and could act as a strong impediment to arms control measures. Restrictions in this area could also lead to an increase in black market dealings; indeed the very presence of such a market testifies to economic and commercial motives.

Some general economic features of the arms market need to be stressed. We have already seen that the demand side of the arms market shows relative stagnation or stability. The principal reason seems to be economic. Faced with a generalized economic crisis, in terms of increasing debt, a collapse of commodity prices, structural adjustments and macroeconomic stabilization policies which call for reduced government expenditure, Third World countries are finding it extremely difficult to import arms. The oil price boom seems to be over and economic stagnation is a general problem even for the richer developing nations. Regional conflicts, though widely prevalent in the 1980s, are generally using less imported arms (for example Iran and Afghanistan as compared to Vietnam).

On the supply side there are essentially six features which encapsulate the economics of arms exports particularly for the developed economies:

- At the domestic industrial level there is increasing concentration with large firms producing a range of very sophisticated products.
- At the international level the number of countries actively participating in exports has increased over the last decade. In addition to the superpowers and the four other "majors", there are currently about 23 countries which regularly sell weapons in the world market.
- Demand stagnation coupled with increasing numbers of supplying countries clearly means that market shares for individual sellers are declining.
- There is more competition at the international level, a reversal of the trend at the domestic level. The index of concentration among the three groups of exporters, shown in Table 6, indicates that concentration is declining as among countries. Competitive tendencies imply that the national buyers are given better deals and more encouragement as well as innovative new sales arrangements involving CTAs.
- The foregoing features in turn imply increasing commercialization with explicit economic motives often dominating political and ideological considerations. Thus, Kuwait and Saudi Arabia propose to buy from the Soviet Union; Israel sells to South Africa; Libya keeps its European options open; China sells to both Iran and Iraq during their long war.
- Technology transfer and proliferation are now an integral part of the weapons trade. This increases significantly the number of Third World producers, who may themselves wish to participate in the trade relations.

All these features are interrelated and produce a symbiotic framework which explains the economic aspects, taken together, of the arms trade quite clearly. At the domestic industrial (firm) level the technological features of production require large-scale investment and the attainment of a huge productive capacity before profits can be realized. The firm or industrial organization, therefore, tends to be "large" to take advantage of scale economies. A few large firms will dominate the market and an oligopolistic structure will prevail. A characteristic feature of oligopoly is the creation of entry barriers such that potential entrants can be barred from starting production and eroding existing profits. An effective entry barrier is to raise the sophistication of the existing products, increase Research and Development (R&D) as well as to produce more advanced quality output. This leads to vertical proliferation with military technology outpacing the needs of military sufficiency. *Invention becomes the mother of necessity.* Few new entrants dare to enter the market since the "price" of entry, in the form of matching up to the existing specifications and bettering them, are prohibitive. The domestic market is characterized by high concentration and a preponderance of advanced technology.

However, these large firms also need high sales to make their production capacity viable. If the domestic market in the form of Defence Ministry purchases are buoyant then there are few problems. The 1980s US defence industry has been characterized by precisely such buoyancy and sales growth, given the unprecedented increase in military spending. It has been estimated that US defence expenditure will be $270 billion *more* under the Ronald Reagan Administration than it would have been if pre-Reagan budgetary trends had been maintained [13]. Under these circumstances there was little economic incentive for exports. The case is radically different for the so-called major commercial exporters, France, UK, FRG and China (see Table 6). Given that security perceptions require a viable domestic arms industry, while at the same time the internal market is too small to absorb full capacity sales, economic motives force export promotion. The increase in global export market share is an indication of how aggressively these countries have tried to promote their sales. However, a fundamental and major constraint in the 1980s is the relative stagnation of aggregate demand. Hence, even as they tried to improve their position they have increasingly found that it is difficult to increase global sales; the aggregate demand curve has refused to budge.

In the scramble to increase demand, firms and governments have taken recourse to more imaginative trade deals. The whole gamut of CTAs has been used to force additional sales [14]. The effect of these is clearly seen in technology transfers and licensed production in the Third World itself, which have boomed. This horizontal proliferation has also increased with more and more countries involved in domestic manufacture of arms. Unfortunately, the vicious circle cannot be stopped. As countries produce they also wish to enter the export market. There is a mistaken belief that exports will make the domestic industry more "profitable". But of course if everyone wishes to create an export "miracle" the final outcome is counter-productive. The end result is that there are more individual countries involved in arms exports, each with smaller shares, while aggregate demand remains stable. The economics of arms trade has become self-defeating.

5. Regional Conflicts

Our previous analysis has shown that the 1970s saw a remarkable growth in the arms trade with the Third World followed by stagnation; in addition, from around 1977 the level of the arms trade has stabilized. To relate these two macrofeatures with regional conflict we need to compare the incidence of such conflicts between the 1970s and the 1980s. For various reasons, this is not an easy task. Firstly, there are difficulties in comparing the intensity of conflicts and resultant wars. Secondly, certain conflicts may remain incipient without much external manifestation such as a battle; yet they may cause arms imports to rise. Thirdly, it may not be possible to distinguish between an internal revolt or civil war, and an external and regional outbreak of hostilities. Fourthly, a particular time period may be characterized by an exceptional conflict which tends to distort the overall picture. The Iran-Iraq war is a specific example of this phenomenon and will, therefore, be dealt with separately.

We must begin with consideration of some of the major regional conflictual hotspots [15]. We will compare the intensity of conflicts, and particularly wars, with the growth rate in arms imports. The latter are to be found in Table 1. In the Middle East the triangular confrontation among Israel, Egypt and Syria (and possibly Jordan) led to more open hostilities during the earlier decade. Border conflicts, following the six-day war (1967) continued into 1970. The Yom Kippur war took place in 1973. Also, Syrian intervention in the Lebanese civil war dates from this period. The current decade has seen some reduction in tension, particularly after the Camp David agreement between Egypt and Israel. Incipient tension continues, however, particularly between Syria and Israel; and the Lebanon war has become much worse. Overall, the fall in arms imports growth as between 1970–79 and 1980–88 can be traced to the reduction in conflict taken as a whole for the region. In addition, for the countries in the Gulf belonging to the Organization of Petroleum Exporting Countries (OPEC), the decline in oil revenue (in real terms) has meant widespread budgetary cuts which have had their effects on military spending.

In the case of South Asia, the situation is reversed. The major players, India and Pakistan, fought a full scale war in 1971; thereafter there have been sporadic and minor clashes. During the 1980s relations between the two countries have even improved slightly. Yet, there has been a rapid spurt in arms acquisition. The fundamental reason seems to relate to Afghanistan, even though neither India nor Pakistan are directly involved, nor can they conceivably think of any specific security threat from the Soviet Union itself. The Soviet involvement in

Afghanistan has been used as a pretext by Pakistan to enlist the help of the United States in promoting re-armament. The ensuing arms race with India has caused an explosive growth of arms imports into the sub-continent. This is the classic example of superpower involvement, either directly or by proxy, which exacerbates an implicit conflict into a more explicit and explosive situation though the supply of arms.

On balance, the Far East has seen less security problems in this decade as compared to the previous one. The Vietnam War, which ended in 1975, required massive importation of arms into the region; so did the US intervention in Laos and Cambodia. In addition, the Sino-Vietnamese conflict in 1979 erupted into a costly border clash. In comparison, the Cambodian conflict (1978–1987) has been of relatively low intensity. The growth of arms imports between 1980 and 1988 is predominantly explained by the economic affluence of the Asian-Pacific countries and their high economic growth rate coupled with an export miracle.

Trends in North African arms imports are clearly explained by the economic fortunes experienced by the countries in the region (particularly Libya and Algeria). The oil price rise transformed their military demands and they rapidly rose to the front ranks of arms buyers. Regional conflicts (Morocco, Chad and Libya) have remained at similar levels throughout the period under consideration. Fluctuations can be precisely attributed to oil revenues.

Economic crisis also explains the fall in Sub-Saharan Africa's growth rate of arms. If anything, the incidence and intensity of war has increased. This is particularly true of Southern Africa where civil and inter-state wars, instigated by South Africa, have threatened the societies of Angola and Mozambique. The front-line states have all been forced to import more arms than would be necessary under more normal times simply because of the destabilization [16]. In spite of this the growth rate has plummeted because nations simply cannot afford to purchase more arms even for their genuine security needs and self-defence. However, there are some individual exceptions. For example, Angola is now black Africa's largest importer of major conventional weapons – a direct effect of war.

The same trend holds true for Central America. Conflicts in El Salvador and Nicaragua, as well as in other countries like Guatemala, are on the rise. But arms purchases have fallen as countries grapple with economic difficulties. A similar story could be told about South America. Here, internal conflicts (for example in Peru and Colombia) have risen dramatically while external or inter-state troubles have diminished (for example, between Argentina and Brazil). But the central reason for restraint in the arms trade must be the debt crisis and the hyper-inflation which forces general austerity on these countries.

In many respects, the Iran-Iraq war is a unique example of Third World conflicts, though some lessons can be learnt. The actual hostilities have continued for the longest period since the Second World War. But in spite of massive attrition and horrible losses suffered by both sides, the economic infrastructure remains basically intact; both nations have benefited from oil revenues, and government budgetary constraints are not too rigid. Even the oil wealth has not been able to compensate, however, for the deprivation and socio-economic problems that the countries have to bear.

Table 7. Iran and Iraq: Value of Arms Exports; Totals for Selected Time Periods (US $ billion at constant 1985 prices) (Bracketed terms are *annual* values)

	Iran	Iraq	Total	Iran's Share
Pre-oil (1971-73)	4102	998	5100	80.4
	(1367)	(333)	(1700)	
Oil boom (1974-78)	16037	4062	20099	79.8
	(3207)	(812)	(4019)	
Pre-war revolutionary	78	2540	2618	3.0
(1979-90)	(39)	(1270)	(1309)	
War (1981-88)	3112	22528	25640	12.1
	(389)	(2816)	(3205)	

Source: SIPRI data base and author's calculations.

Table 7 summarizes the data calculated from Brzoska and Ohlson [17], and SIPRI [5]. Arms imports, between 1970 and 1988, are categorized according to major time periods; pre-oil price rise; the oil boom; the Iranian revolutionary period just before the outbreak of hostilities; and finally, the war phase itself. An interesting pattern emerges and the conclusions will be summed up here.

- During the Shah's reign Iran went on an arms spending spree which led to its regional share becoming around 80 per cent of the total. Seeking to act as the policeman of the Gulf, Iran dominated the weapons import league in the region.
- The revolutionary Iranian regime eschewed foreign arms and prior to the war had dramatically reduced their country's share to a negligible 3 per cent.
- Even during the war itself Iran was a relatively small purchaser of arms with a share hovering around 12 per cent. This was partly due to the embargo and the alienation of the superpowers. But it was also because the authorities believed in self-sufficiency rather than external dependence, as well as trusting in ideological fervour rather than modern equipment as a means of conducting a war.
- Iraq's war preparation, in the face of perceived weakness and the disorganization of its traditional rival, was clearly evident in the period 1979–80 when there was an overshooting of both the level and the share of its weapons imports. This phase continued into the war period when Iraq relied more heavily on arms importation to substitute for weaknesses in manpower and strategic geographical depth. Effectively, it utilized an equipment-intensive war-fighting technology in line with its comparative advantage.

Are there major lessons to be learnt about the relation between war (explicit regional conflict) and arms imports from this brief case study? A spurt of imported arms acquisition by one country, in a region with a history of implicit conflict and rivalry, can be destabilizing. This may not be apparent during the period itself since the relevant country (Iran under the Shah) could appear to have regional dominance ("the policeman"). However, the inherent risks never disappear. When there is a structural or political shift (the Iranian revolution) the formerly weaker countries may wish to take advantage of the situation. The result may be war, as in this case. The implications for other regional rivals, such as India and Pakistan, are obvious.

6. Conclusion

There are two principal reasons why Third World nations import arms and related technology, particularly major weapons systems. Regional conflicts and related wars are a major impetus to importation since such destructive assets are subject to attrition losses and rapid obsolescence. However, an equally important reason seems to be in economic prosperity and a rise in the wealth of nations. The latter is particularly important when there is a rapid rise in wealth, for example, after the oil or commodity price boom of the 1970s. Conversely, when there is a developmental crisis forced by falling export prices and rising debt burdens then the import of arms fall. The combination of conflict, economics and security, therefore, produces a complex picture.

The relative stability of Third World arms imports is set to continue mainly because of the economic crisis facing such countries and the bleak prospects for economic growth in the near future. But the level is still high even though heavily concentrated among about 15 nations. Arms control measures are, therefore, crucial in the field of weapons trade [18]. Sadly, the prospects are not encouraging given qualitative proliferation, technology transfer, the slow rise of the black market and the militarization of Third World conflicts. In addition, the growing role of commercial factors are a cause of some concern since the political will may be diluted in the face of increasing pressures from industries to sell more abroad. There may even be a contradiction within the domain of East-West arms control which reduces the domestic demand for weapons procurement among developed countries; the pressure to diversify abroad may be irresistible. Policies for industrial conversion are, therefore, an essential part of the general process towards disarmament. There is now a strong need for policy coordination between the North-South and East-West dimensions of arms production, procurement, trade and controls.

Appendix

There are three ways of calculating the per annum growth rate of a time series. The first is to take the proportional rate of change on an annual basis utilizing the initial and latest values of the series. The problem with this method is that it solely depends on two extreme values only and, therefore, can neglect the intermediate terms. The second way is to take year-to-year proportionate changes and make a simple average of them. The difficulty with this method is that it can give too much emphasis to individual fluctuations; any outlier will have an undue effect on the average if it is large enough. This is particularly problematic when the relevant series is itself prone to rapid changes; the arms trade is a good example since it fluctuates widely. The third method is to take a least squares regression of a semi-log linear specification of the relevant variable with respect to time. This method accounts for all the variables in the series, as well as giving less importance to outliers. One problem is that a growth rate can be quite large and yet be statistically insignificant (acceptance of the null hypothesis that the coefficient is not significantly different from zero). The t-statistic has been checked and wherever insignificant the relevant growth rate has been asterisked(*) in the Tables. From a statistical point of view these growth rates should be considered as zero.

For any variable x_t the exponential growth formula is

$$x_t = A \exp(gt)$$

where g is the growth rate and t is time. The regression equation used to derive the relevant value is:

$$\log x_t = \log A + gt \quad .$$

$\log x_t$ was regressed on t (time) and the coefficient g is shown in the Tables.

The index of market concentration used in Table 6 is the inverse of the sum of squared market shares (the "Numbers Equivalent Herfindal" (NEH). If there is no concentration – each group has a share of one-third of the market – then the maximum value that the index can take is 3. When the index rises, as between the 1970s and 1980s in Table 6, concentration is clearly going down.

References

1 US Arms Control and Disarmament Agency, *World Military Expenditures and Arms Transfers, 1987*, (Washington, DC, 1987)

2 See Stockholm International Peace Research Institute (SIPRI), *World Armaments and Disarmament: SIPRI Yearbook, 1989* (Oxford, 1989)

3 Derived from Office of the [US] Assistant Secretary of Defense (Comptroller), *National Defense Budget*. Estimates from 1988/89 (Washington, DC, 1988)

4 World Bank, *World Development Report, 1987*, (New York, 1987)

5 See SIPRI, *The Iraq-Iran War 1980–1988*, (Stockholm, 1988)

6 A. J. Pierre, *The Global Politics of Arms Sales*, (Princeton, New Jersey, 1982)

7 The Stockholm Peace Research Institute, *World Armaments and Disarmament: SIPRI Yearbook 1982*, (Taylor and Frances, 1982)

8 S. Deger, "The Economics of Soviet Arms Trade", in R. Cassen (ed), *Soviet Interest in the Third World*, (London, 1985)

9 S. Deger, *Military Expenditure in Third World Countries*, (London, 1986)

10 See I. Anthony, "The Trade in Major Conventional Weapons", in SIPRI, *SIPRI Yearbook 1989*

11 E. A. Kolodziel, "French Arms Trade: The Economic Determinants", in SIPRI, *World Armaments and Disarmament: SIPRI Yearbook, 1983*, (London 1983. See also E. A. Kolodzief, *Making and Marketing Arms: The French Experience and its Implications for the International System* (Princeton, New Jersey, 1987).

12 U. Albrecht, "The Federal Republic of Germany and Italy: The New Strategies of Mid-sized Weapon Exporters?" *Journal of International Affairs*, vol. 40, no. 1 (1986–87)

13 S. Deger, "World Military Expenditure", in SIPRI, *SIPRI Yearbook*, 1989

14 S. Neuman, "Coproduction, Barter and Countertrade: Offsets in the International Arms Market", *Orbis*, vol. 29, no. 1 (1985)

15 Information derived from publications over several years by Ruth Sivard, *World Military and Social Expenditures*

16 UNICEF, *Children on the Front Line*, (New York, 1987)

17 M. Brzoska and T. Ohlson, *Arms Transfers to the Third World, 1971–85*

18 H. Wulf, "Arms Transfer Controls: the Feasibility and the Obstacles", in S. Deger and R. West (eds), *Defence Security and Development*, (London, 1987)

The Arms Trade and Regional Conflict: Suppliers' Policies and Behaviour and Their Consequences

Abdel Monem Said Aly

For much of the post-1945 era, arms transfers have been a crucial part of the major powers' relations with the Third World. Although these powers have tried to establish close ties with the developing countries through a variety of non-military methods – including economic assistance, trade, educational exchanges, and political cooperation – arms transfers have remained the major instrument in building and changing regional alliances. Thus, in spite of the centrality of East-West relations in world politics, the Third World has been the field where indirect confrontation has taken place. Of course, regional factors, historical as well as geopolitical, have played their part in leading to the great powers being called upon for help. However, Third World conflicts presented an opportunity for the major powers, and especially the superpowers, to intervene directly and indirectly in these conflicts. In the last four decades certain patterns for arms transfers have emerged.

Firstly, the market for major items of military equipment has been highly concentrated. The five largest exporters – the Soviet Union, the United States, France, UK and China – have provided over 80 per cent of permanent seats on the UN Security Council. Theoretically, according to the UN charter, they are responsible for international peace and security. Although they have succeeded in maintaining peace among themselves due to their nuclear deterrence capabilities, they have failed to maintain peace in the Third World. Regional wars have been a major aspect of the history of the post-war era. The five permanent Security Council members were often involved in these conflicts directly, as in the case of Suez, Vietnam, and Afghanistan, or indirectly through arms transfers. On certain occasions, as in Cuba in 1962 and in the Middle East in 1956 and 1973, their involvement could have led to global confrontations.

Secondly, among the five leading arms-exporting countries, the United States was the leading exporter until the late 1970s when the Soviet Union became the leading supplier to the Third World. This change came as a result of the US military build-up during the 1980s on one hand, and on the other hand, the decline in Soviet reliance on non-military means of achieving foreign policy objectives. For the Soviet Union arms transfers have become the primary instrument enabling it to compete with Western countries and China for influence in the Third World. Additional factors have consolidated this prominent Soviet position:

- The Soviet Union has a decision-making process that facilitates the speedy delivery of arms. While the Soviet Union usually takes less than 12 months from contract to delivery of major weapons, the United States takes from 24 to 36 months or longer. This delay in the US delivery system is due to the pressures from public opinion (particularly in the case of delivery to Central America and the Middle East) and the scrutiny of the State Department, Defense Department and the Arms Control and Disarmament Agency. In addition, there is often a lengthy congressional review; and in certain cases (Peru, Jordan, and Saudi Arabia) Congress has been able to veto arms sales.
- The Soviet Union has a large stockpile of surplus equipment ready for export once the political decision has been made. The availability of surplus equipment allows Moscow to export without having to deprive the Soviet Armed Forces of new procurement. Western countries tend to rely on equipment that is currently operational, for example, when supplying Israel with weaponry for 1973.
- Soviet weapons have been more "economical" for many Third World countries. In addition to supplying arms sometimes free of charge, Moscow has offered her customers lucrative concessionary sales and soft terms of financing. Furthermore, the Soviet Union, unlike many Western countries, has been very adept at developing barter arrangements. This has been more appropriate to the circumstances of Third World customers who suffer from extreme scarcity of hard currency revenues.
- Soviet weapons have proved more suitable for Third World use. They are easy to maintain and operate. This has been an important advantage for Third World countries, whose armies consist of illiterate conscripts lacking technical skills and suffer from shortages of technicians.

A third feature of the recent arms trade market has been that it is largely motivated by political and military concerns. The Soviets, Americans, French, British and Chinese have used arms exports to advance political and military objectives. Soviet exports to Egypt in the mid-1950s, to Cuba in the 1960s, to Peru in the 1970s, and to Nicaragua in the 1980s, enhanced Soviet positions in areas traditionally dominated by the West. Massive American arms supplies to Israel during and after the 1973 War were to balance similar supplies to Egypt and Syria. By the late 1970s and during the 1980s, American, French and British supplies to the Middle East were instrumental in the de-radicalization of the area, limiting Soviet influence to a small number of states (notably Syria, Ethiopia, Libya and South Yemen), and increasing Western influence at the same time. The same is true of Soviet arms transfers to India, which helped Moscow to gain influence with the Indian government and to balance American support for Pakistan. Another important goal of arms transfers has been for the supplier to obtain overflight and landing rights, port facilities, bases and prepositioning of equipment. As a result of arms transfers, currently in the Middle East the United States has naval and intelligence facilities in Bahrain, use of airfields in Egypt, use of airfields, naval communications and storage facilities in Morocco and Oman, and the use of naval facilities in the United Arab Emirates. The Soviet Union has the right to use airfields and naval facilities in Libya, South Yemen, Ethiopia and Syria. Moreover, arms transfers, particularly in tense Third World situations, help to test and gain valuable information about the combat performance of weapons. The Middle East with its major regional conflicts (the Arab-Israeli conflict, the Iran-Iraq War, and the conflicts in the Horn of Africa and in Afghanistan) has provided extensive data about the performance of major weapon systems produced by major supplier countries. Finally, arms transfers have been a valuable means for arms-producing states to conduct espionage in the recipient countries and also indirectly against adversaries through the dispatch of advisers and technicians accompanying the weapons. The growing linkages between the supplier country and the military institutions of the recipient countries demonstrate the tendency to mould the latter politically and strategically in the image of the former. Soviet "advisers" helped shape the policies of the Kabul regime after the Communist coup of April 1978. Likewise in South Yemen in early 1986 the presence of Soviet military personnel enabled the Soviet Union to regain control of events after bloody fighting erupted between rival factions of the ruling Marxist party. Other examples can be seen in South Vietnam and Iran where US advisers played similar roles.

A fourth major aspect of the recent arms trade market is that if the supply side of arms trade in major weapons has been highly concentrated in a few highly industrialized major powers, the recipient side has been equally concentrated. About 15 countries of the Third World have dominated the importation of arms. All of them are involved in long-standing, intense and major regional conflicts. India, Iraq, Egypt, Saudia Arabia, Israel and Syria are at the top, not only of the Third World, but also at the top of world importers. These countries have received top quality weapons from the arsenals of major suppliers. This has contributed to the intensification of the regional conflicts in which these countries are involved. The Indian-Pakistani conflict, the Arab-Israeli conflict, and the Iran-Iraq conflict would not have continued with the same intensity without the availability of sophisticated arms provided by major suppliers.

The previous patterns just described concern the general characteristics of the arms suppliers and their behaviour as they have evolved in the post-war era; there are a few countries at the top of the hierarchy of power in the international system; they are politically motivated, and they tend to concentrate their supplies on a small number of recipients involved in highly explosive regional conflicts. But in the 1980s a few new patterns have begun to emerge. And it is to these that we must now turn.

First, beginning in 1983, the total value of arms trade for major conventional weapons has been declining, with the year 1987 being a sole exception. The decline can be attributed to the following factors:

- The Third World debt crisis has been a feature of the 1980s. And the principal Third World arms recipients are also ranked, with the exception of Saudi Arabia, among the top debtor countries.

- The debt crisis has been coupled with the decline in the price of Third World raw materials, particularly oil. This particularly curtailed the ability of the Arab oil–producing countries to purchase weapons. The oil returns of Arab members of the Organization of Petroleum Exporting Countries (OPEC) were a mere $2.169 billion in 1965. By 1970, returns reached $51.5 billion and in 1980 they sky-rocketed to $204.244 billion. Since then, however, oil revenues have declined annually to $182.9 billion in 1981, to $134.7 billion in 1982, to $101.7 billion in 1983 and to $102.4 billion in 1984. The 1989 returns ar not expected to be more than $67.7 billion. Thus, in its rise and fall, oil prices have been one of the determining factors for arms trades.
- It has become increasingly difficult for certain countries to absorb more weapons. For example, Saudi Arabia for most of the decade has been struggling to integrate its previous arms purchases into its force structure. And Libya has been saturated with arms to the extent of being incapable of manning them.

A second new pattern is that although the arms trade has continued to be politically-motivated, economic factors have tended to be more influential in the 1980s. As the American budget deficit climbed during the decade, arms exports became one of the means of reducing it and of contributing positively to the balance of payments and trade. Of the $128 million that the Saudis were paying for each Airborne Warning and Control System (AWACS) plane, about $60 million went to the US Treasury as partial repayment for government Research and Development (R & D) expenditure. This has become true of all high-technology military equipment sold by the United States. The economic stakes for weapon-producing companies are likewise a factor in the arms trade. In all Western countries military industrialists have been lobbying their governments to ease the way for their arms exports. When the US Congress tried to restore its power over arms exports through the Biden-Leving amendment to the 1976 Arms Export Act, American industry proved capable of removing the bill from the Congressional agenda. Many legislators were convinced by arguments from industry that it would create an unfair disadvantage for US arms exporters.

Economic considerations are more influential in Socialist countries where the economic crisis is more evident. From the early 1970s on, the Soviet Union started to experience hard-currency trade deficits when it had to import large quantities of grain to make up for a series of crop failures. Consequently, Moscow started to look to the arms trade as a source for hard currency earnings. In the early 1970s, arms transfers provided only 3 per cent of Soviet hard currency revenues; by the end of the decade, they accounted for 10–15 per cent of such earnings. With the decline in world commodity prices in the mid-1980s, the share of hard currency that arms sales contributed to the Soviet treasury has risen to 25–30 per cent. Arms transfers also benefit the Soviet economy by maintaining excess capacity in Soviet defence industries at virtually no extra cost. In addition, as is the case in the United States, arms transfers help spread the burden of Soviet R & D costs for particular weapons.

China provides a clear example of the rise of "economism" in the arms trade world. Since the late 1970s, it has become a major arms supplier by being the only state which supplied major weapons to the conflicting parties in the Iran-Iraq War. Economic factors were behind this behaviour. Similarly, the same factors can explain the Polish and Romanian arm supplies to the Nicaraguan Contras.

A third new factor witnessed in the 1980s was the rise of a few Third World countries on the list of arms suppliers. Brazil, Israel and Egypt are gaining ground in the world arms trade. At an intermediary technological level Third World producers have captured a considerable share of the international market. Between 1974 and 1985, Third World arms exports rose from about $1.5 billion to $8.5 billion. This rise in the Third World role in the world arms trade has been enhanced by cooperation agreements among Third World arms producers. Brazil has agreements with Argentina, Egypt and Nigeria. Egypt has another with Turkey. These agreements tend to reinforce the "South-South" trade patterns which have evolved over the past decade. Furthermore, in order to enhance their technological capabilities a few Third World countries, notably Israel, Egypt and Pakistan, have attempted to steal sensitive designs and materials helpful to their arms industries.

The 1980s have also witnessed the rise of the grey and black markets in arms transfers. This is hardly new in the arms trade. However, in the post-war era, the arms trade increasingly became a matter of public policy controlled by national governments. The Iran-Iraq War which dominated the 1980s encouraged both the grey and black markets. By the second half of the 1980s, most of the major arms suppliers (the superpowers, UK, France, China, Argentina, Israel, Libya, North Korea, Portugal and Spain) were involved in grey-marketing arms. The grey market includes officially approved exports from governments which do not want to be associated with their own actions. The black market, on the other hand, includes non-officially approved exports and is handled through private networks. Arms supplies to Iran and South Africa have been provided through the black market. More importantly, however, the black market has become the arena for supplying knowledge, materials and instruments for Third World countries enabling them to build capabilities for mass destruction weapons: nuclear, chemical and biological. It has been reported that West German and Swiss firms have been embroiled in assisting Iraq, Libya, Egypt and Syria to build such deterrence capabilities.

A further feature of the 1980s has been that in order to keep their share in the arms trade, major suppliers have resorted to counter-trade offset and co-production agreements. As has been mentioned previously, this had been common practice for the Soviet Union since the 1950s when it provided arms in exchange for Egyptian cotton. But by the end of the 1980s other countries had followed. The United States and France have coproduction agreements with Egypt and Israel. Washington has signed *strategic cooperation* agreements which give Egypt and Israel a non-NATO ally status. Under these agreements, both countries are allowed to participate in specific joint R & D projects with the United States.

Great Britain's deal involving Tornado aircraft for Saudi Arabian oil is one of the largest counter-trade deals in modern history. Of course, the Soviet Union has continued these practices particularly with India. Moscow signed a co-production agreement with Delhi for the Mig-29. This trend enables cash-poor countries to overcome financial difficulties and remain dependent on their suppliers.

The prospects for arms trade in the 1990s are not easy to define. Most likely they will be partially conditioned by the trends of the 1980s, as well as by the new circumstances that will emerge. Some speculations are in order:

- If the reconciliation in Soviet-American relations continues along the lines seen in the second half of the 1980s, it is possible that the polarization effect of regional politics will also decline. Thus, regional conflicts will tend to decline and the need for arms will decrease.
- We have recently witnessed a trend for regional settlements in Central America, Southern Africa, the Middle East, South Asia and Southeast Asia. If this trend continues, there are greater possibilities for a serious decline in the arms trade.
- Among the regional conflicts, however, the Arab-Israeli one and the Central American protracted conflicts seem the most difficult to resolve. A continuation of tensions in these areas might lead to a major war at a much higher level of armaments both quantitatively and qualitatively. These two areas, particularly the Middle East, might contribute to an upsurge in arms acquisitions in major systems.
- In addition to these two regional conflicts, it is possible that the decade of the 1990s will be the decade of civil wars. The decline of the economic fortunes of many Third World states is contributing to the rise of social and political tensions in these countries particularly in Sub-Saharan Africa. This might enhance the prospects for trade in small arms and less sophisticated equipment which Third World producers and Socialist countries are ready to supply. The Lebanese and Sudanese models probably will characterize the war mood of the coming decade.

Socio-Economic Consequences of Arms Transfers

Virginia Gamba

Arms do not necessarily lead to conflict. They do, however, have an influence on the level of violence applied to the solution of a conflict. Again, this is not necessarily true of all types of weapons. For example, many thinkers have built their deterrence theories on the existence of a vast number of nuclear weapons which – they hope – will never be used precisely because of the utter destruction this mass of weapons could bring about. But only the superpowers have actually acquired nuclear weapons on this scale. The fact is, moreover, that theories of absolute deterrence based on nuclear weapons alone have not proven enough. Over the last thirty years we have seen a change in the pattern of military doctrines relating to the potential use of these weapons in war. By and by, nuclear weapons have come to be recognized, if not as a useless asset, then as very much an incredible one. Thus, new theories linking the possession of these weapons with the capabilities of more deadly and modern conventional weapons have emerged.

When we talk of "deterrence and defence", of "limited wars", of "flexible response" and even of "non-offensive defence", we are talking about power deriving from a spiralling technological race in conventional weapons. Since nuclear-weapon states consider their nuclear arms largely as a political asset – a perceptual matter, conventional weapons have again become the real issue in the power considerations among nations. Though minimized in writing when compared to nuclear weapons, they are in fact used to give credibility to these same weapons. More importantly, they are used to enable a nation to use force when it feels it cannot settle a dispute by other means. Modern conventional weapon capabilities allow countries to fight. The possession of such weapons in vast quantities and types is also seen as enabling a nation to fight "proportionately", according to the objective, the enemy and the stakes in any one conflict, be it ideological, military, economic or political. But weapons do not make conflicts, men do.

Weapons are a vehicle for men to negotiate the temporary solution of a conflict to their advantage. By the same token, the possession of large quantities of conventional weapons of limited life-span – since technological progress in this field does not allow countries on a once-in-a-lifetime basis to establish a permanent stockpile – makes the possibility of use of such weapons more attractive and "thinkable" to the nations who possess them.

For a long time, the possession of weapons was tied to the notion of offence or attack, but with the need to contain violence, especially after two world wars, men devised a different notion for the possession and use of weapons. Now they are tied to the notion of defence. Thus, it is increasingly difficult for governments anywhere in the world to abstain from the possession of weapons. Indeed, many a state hoards weapons under the principle of "maintaining peace". When this occurs, the production and the bartering of weapons between nations comes to figure largely as a percentage in a national budget.

In peace, weapons have a tendency to "disappear from view", that is they become intangible. They are open to criticism only when their numbers, represented in a budget, are thought to affect other budget commitments. In peace, many governments take pains to minimize the visual impact of the weapons they are buying and producing. When approval of a budget is necessary, they rely on convincing the political representatives that they are in fact paying an insurance premium to protect and retain peace. These weapons, they argue, will not be used precisely because they exist.

What really demonstrates the duplicity behind the possession of weapons is the testing of the logic in times of war. During war the weapons become tangible, they become visible once again and, more often than not, they will be judged inadequate and pitifully less powerful than expected. Then the cries will be for more since what is there is not enough. This phenomenon will be much more marked if the country in question is fighting a losing war.

Thus, possession of vast amounts of conventional weapons will be judged negatively during times of peace and as inadequate at times of violent conflict. More importantly, there is a fine line in the arms business. Possession of too many weapons might cause pressure from below to use them at the first available opportunity. Should this not occur, then the pressure from below turns on the decision-makers questioning their amassed, unused, military power. This leads us back to the original question: is the possession of weapons a generator of conflict?

Conflict exists before weapons appear. It is perhaps inevitable that nations should have conflicts of interest between one another. After all, nations are living entities that grow and expand according to the principles of their own life. Sooner or later their interests, when projected, will clash with those of another country, even turning friends into foes and vice versa. These conflicts are not only military. They often start out as something other than military; on many occasions economic, political, cultural, religious and diplomatic conflicts are the origins of a military response.

Weapons do not create conflict. More importantly, weapons cannot solve a conflict unless they are accompanied by a series of far-reaching socio-economic, diplomatic and cultural alternatives that will change the *status quo* of the original conflict into something other than a dispute. Let us consider the interaction between nations at different levels of violence. When two nations are "at peace" in reality they are in a standoff situation in which both agree that it is better to negotiate their differences peacefully. They have both reached a *status quo* that allows for their differences to be contained. But when something occurs to change this *status quo*, and one of the nations perceives that containing a dispute at a given level is no longer acceptable, then a crisis situation sets in where one side will strive to convince the other either to return to a previous state of affairs or to accept a new situation that favours one over the other. Not surprisingly, a crisis situation often escalates to all-out war. But when this happens, both sides will often initially attempt somewhat to limit the original thrust of the warfare, so as to retain the capabilities to escalate to an all-or-nothing effort.

It is the capability that is retained, not the one actually used at the beginning of a military conflict, which may enable a peaceful solution to the dispute to be negotiated even though violence has already started. This simple logic underpins the theory of both nuclear and conventional deterrence. The problem is that, since deterrence does not end with the initial crisis or with limited war, though it may be considerably weakened, the perceived need to hold on to a credible threat to engage in further escalation leads nations to over-insure militarily in time of peace. Thus, the old identifying line between a country's intentions and its capabilities is lost. For today a country that is rapidly militarizing and spending vast sums of money on military *materiel* is not automatically judged to harbour aggressive aspirations towards its neighbours. Everything is permitted under the mantle of "peace".

The various concepts which do and do not make sense are the ones that are propagated by the countries that today occupy the centre of the international stage, both in military and economic terms. But the vertical arms race which they have started with their logic is now spreading horizontally to other states. Nations are forced to militarize, either because they form part of an alliance that requires them to do so regardless of the real threats to their actual national interest, or because they face such an alliance. By the same token, less developed nations which hope to enter the international discourse through economic or cultural projections, will inevitably fall prey to pressure to accept standardization, not only of international finance, weights, measures and trade, but of military doctrine and systems.

It has often been said that military power is a communicator between countries. This occurs because, despite language and cultural barriers, everyone understands the meaning of a tank, a fighting aircraft, a warship or a man with a gun. The symbol transcends the barrier of differences. Thus, Third World nations understand and take on the values that industrialized nations defend for their own use. More importantly, since the basic problem of the Third World is one of underdevelopment and of economic struggle, more often than not the question of arms will be tied up with other international deals. At times they become the *quid pro quo* in such a relationship. Thus it is easy to see economic aid from a developed nation to a lesser developed one culminating in an exchange of weapons. Bilateral trade and economic investment are often followed by military assistance, weapon procurement aid or arms transfers. The same is true the other way around: those Third World countries which have their own embryonic military-industrial complex will attempt to find markets for their produce among more developed countries, which must provide this outlet if they aim to enter the market of the former at other levels. If the military-industrial complex finds no outlet, then it can become a bargaining chip for other types of relationship with similar nations or with even more less-developed ones. Thus, a commercial agreement between a country of the Third World and another, even less developed than the first one, might be decided on the potential military aid and sales that the first one can induce the latter to accept. Examples abound where arms transfers are not the main issue but the symbiotic issue by means of which one nation strives to gain a foothold of influence over another. And this occurs mostly because of growing dependency once arms transfers have taken place.

The issue of arms transfer is a complex one for these and many more reasons. But, without a good understanding of the socio-economic consequences of arms transfers, it cannot be approached. In this sense, I believe that socio-economic consequences of arms transfer are two-fold. They refer to the countries that produce and transfer arms and to those that receive them.

For the producer nation, the issue of arms transfer has a very important socio-economic consequence. The decision-making body in such a nation will be able to allay the pressure from below their own establishment *vis-á-vis* their arms production because they can demonstrate that they are increasing the national product by so doing. By the same token, they will strive to demonstrate that their arms sales policy increases exponentially the influence they retain over potential buyers, and thus opens the door for further bilateral business and investment. It follows, that for many of these decision-makers, the issue of arms transfer is seen almost as an exercise in "confidence-building measures" that will tie-down the recipient country to their produce and influence, thus resisting the temptation to overtly follow policies that might have a negative effect on the national-interest of the producing nation.

Despite this logic, the reality is not so clear. When a nation sustains this policy with arms transfer, especially when the recipient country is economically or politically important to the former, individual issues or crises arising at any one given time will not be dealt with on the basis of their significance then and there. The temptation will be to follow the arms transfer policy even if the particular timing seems negative because too much effort, prestige and money was tied into a given exercise. In short, the freedom of nations to analyse particular circumstances with pragmatism is often curtailed and eroded.

Although, in my opinion, this is the worst socio-economic consequence of the policy of promoting arms transfer, there are certainly others. For example, the military-industrial complex of the producing country can become tied to the client needs and start producing war *materiel* that it has no real use for, since it is obsolete *vis-á-vis* its own natural foes, or because it is impractical for the kind of scenarios it envisages for itself. Thus, a whole gamut of war production will reflect not the surplus of a nation's war fighting capabilities but the careful catering for somebody else's needs. If we add to these negative effects of arms transfers in a producing nation, the temptation to tie arms to political manoeuvering and the ensuing loss of innocence of the population at large, when this is understood the socio-economic consequences are even greater.

At the other end of the spectrum, in the realm of arms recipients, we find another host of socio-economic consequences which prove negative. The recipient country of arms transfers is mostly one that cannot produce its own range of military *materiel* on its own. Perhaps it has an embryonic capacity to do so, or even produce one type of equipment on its own but does not have the capacity to produce the whole range. It follows that it is not a country that has enormous economic strength. More often than not, the countries that buy most weapons are not the richest. More often than not, they are controlled by minorities, elites or the military themselves. Thus, they can allocate to the procurement of weapons a large percentage of their national budget regardless of the socio-economic situation of the country. Since 5 per cent of the Gross National Product (GNP) is a considerable amount of money for a developed nation to spend on defence, it follows that the same amount in a less-developed nation would not cover the same quantities of *materiel.* It is, therefore, not surprising to see some lesser developed nations allocating 30 per cent or more to their GNP for military procurement. The corresponding curtailment of other items that might benefit the population directly is obvious. But, should we claim that a reduction in the size of military budgets in these countries would restore a balance in favour of the social needs of the population, we would not necessarily be stating a truth. Given the type of government I have described, it does not follow that such savings would be spent on the people. More importantly, military might is something that any government can publicly justify as being of 'direct" benefit to the population. After all, it is for *their* defence. This would be the political statement accompanying such investment. In many lesser-developed societies, ignorance and lack of pragmatic and neutral information does not allow the masses to understand what is at stake. Here, manipulation of budgets for military purposes is easier than in other societies. In military prowess and might, a contemporary parallel to the old Roman "Panem et Circenses" occurs, except that at times military might is utilized against the people themselves, as in the case of civil wars.

Another element in weapon procurement and arms transfer concerns timing. Governments in the Third World are often more short-lived than the weapons they order and buy. The procurement ordered by one government is frequently received and paid for by another, at times the direct antagonist of the former. Very few countries have repudiated the arms deals of previous governments. An intangible element of pride comes to the forefront, where the masses are shown the might of their nation through tangible means, and – in these cases – nothing is more effective than an impressive military parade. In this instance, arms transfers offer the possibility of instant demonstration of success, in that, unlike those countries that must rely on their own production of weapons, they can possess modern systems almost immediately. Thus, the might these weapons repre- sent is superficial but showy. The direct consequence of this, however, is that when power amasses, sooner or later, the need to search for a good opportunity to brandish or use it emerges. Violent solutions to historic conflicts are suddenly possible. Yet, when external conflict arises the people usually suffer but only rarely is that suffering related to the notion that military power existed. Rather, it is argued, that there was insufficient to ensure quick victory.

It would be possible to argue that there is a direct relation between arms transfers and conflict or between high military budgets and poverty; but it would not be true. Weapons do not produce conflict, nor is a high military budget responsible for misery in a lesser-developed nation. Poverty and conflict are not the socio-economic consequences of arms transfers and militarization. Many times they are the cause. Nevertheless, there are some direct consequences to arms transfer policies and these affect both the producing and recipient states: in both cases superficial matters will tend to cloud the judgement of decision-makers and act against the pragmatic solution of an endemic problem. In both cases, judgement and the responsible use of political alternatives are impaired when what lies behind arms transfers comes to transcend the original justification.

Control of the Arms Trade as a Contribution to Conflict Prevention

Jasjit Singh

1. Introduction

The period since the Second World War has witnessed nearly 200 inter- and intra-state conflicts. Although many of these conflicts have involved a number of countries beyond the direct participants, and located far from the region in which the conflicts were taking place, all of them have tended to be termed as regional conflicts. Some of these involved direct participation by the great powers and the superpowers, as was the case in Korea, Vietnam and Afghanistan. In nearly two-thirds of these conflicts direct external intervention was an important factor.

Almost all these conflicts are "regional" only in the sense that they have occurred and continue to occur in certain regions; but the near-global linkage is an important characteristic, especially when one focuses on arms supplies from other countries. One significant fact that stands out, however, is that nearly all these conflicts took place outside the continents of Europe and North America, essentially in the territories of the developing world. Since the overwhelming majority of the developing countries, by definition, have limited or no indigenous capabilities for arms production, and thus are heavily – and in most cases, almost totally – dependent on arms supplies from the developed world, it is but natural to focus on the arms trade and its contribution to these regional conflicts. In recent years, an increasing number of developing countries have been increasing their indigenous arms production capabilities. This is a reflection of their progress towards a greater technologically/industrially developed status, as well as of their attempts at increasing autonomy in a critical area. Some of these countries have entered the international arms trade market as suppliers, the more significant among them being China, Brazil and Israel. However, the share of the developing world in arms exports is still marginal – a mere 3.2 per cent, excluding China.

2. Conceptual and Structural Problems

There is both a conceptual and a structural problem involved in any examination of the arms trade and regional conflicts. The arms trade, that is arms procurement through imports, creates, supports and supplements military capabilities. And it is really the military capabilities and the military power of a state that have relevance to the conflicts in which it may get involved. The relationship, therefore, that needs attention is that between *military power* (and its build-up) and regional conflicts, rather than that between the arms trade and regional conflicts. The latter represents only a part of the total and real picture. Indigenous arms production is likely to increase in the developing countries in the future as they move along the developmental path. This would erode the relevance of the arms trade as an indicator of the military power of a state. The conceptual problem may best be seen from the example of Syria and Israel, two states involved in regional conflict for many years. Between 1971 and 1985, Syria accounted for 7.2 per cent of Third World arms imports, while Israel accounted for 5.4 per cent. However, Syria has virtually no indigenous arms production capabilities, not even under licence. Israel, on the other hand, has a vigorous indigenous arms industry which has been exporting arms and military technology to many countries.

The structural problem relates to examining the arms trade in terms of money values. Besides the problems of inaccuracies and inadequacy of data, money value reflects the relationship with military capabilities only indirectly. In most cases, the arms trade reflects only the processes of equipment replacement as an inevitable technical and/or technological imperative. Every successive equipment replacement represents greater technological sophistication and consequently increasing costs. The normal methodology of using cost deflators to reduce the amounts to constant money values helps us to arrive at more representative comparative values; but they fail to address the cost escalation due to the increased sophistication and concomitant effectiveness of weapon systems. Similarly, modern weapon systems also involve much higher components of ancillary equipment in the shape of combat support systems, including command, control, communications and intelligence infrastructures. Degrees of investments in these are not necessarily in proportion to the main combat system; and the overall effectiveness and contribution of the arms trade to that effectiveness remains full of uncertainties.

The arms trade reflects one element of the arms procurement process. However, the scale of arms procurement is very much higher in the developed countries than in the developing countries. For example, the United States alone spends between 80 and 90 billion dollars per year on arms procurement for its military forces; that is, a figure nearly three times the *total* arms trade to the Third World. Europe continues to be the world's most heavily militarized region. And if the argument that this high level of military (including nuclear) capabilities has been the prime reason for Europe having remained at peace and without interstate conflicts since 1945, then the logical corollary is that heavy stockpiles of various categories of arms prevent conflicts; a proposition which totally negates control of the arms trade as a potential contribution towards prevention of conflicts.

3. Need for Arms Trade Control

This brief resumé is not intended to imply that the arms trade and its control has no relevance to regional conflicts. However, it does indicate the constraints and limitations of this approach. The task is not made any easier by the "grey" and "black" market in arms for developing countries which, according to one estimate, may be as high as 50 per cent of the official arms exports to developing countries. Nevertheless, no avenue for preventing or containing armed conflicts can possibly be ignored. And controlling the arms trade does appear to contain desired possibilities and potentialities. A deeper analysis is, therefore, required so that possible measures can be identified.

In order to search for ways and means for the prevention of conflicts, it is first necessary to identify, at least broadly, the causes of armed conflicts. Armed conflicts may be seen in terms of three categories:

- inter-state armed conflicts leading to wars;
- inter-state armed conflicts with use of force not necessarily leading to war; and
- intra-state armed conflicts.

Of these, the first two may be considered together for the purposes of examining their causes.

4. Causes of Inter-state Conflicts

The causes of inter-state conflicts and wars are unquestionably multiple, complex and woven into all aspects of society. Great thinkers, social scientists and strategists have devoted serious attention to identifying the causes and motives of wars. In the 19th century, Antoine Henri Jomini identified the causes of war as follows:

- to reclaim rights or to defend them;
- to protect and maintain the major interests of the state;
- to maintain balance of power;
- to propagate political or religious theories and ideologies, to crush them, or to defend them;
- to increase the influence and power of the state (especially through acquisition of territory); and
- to gratify a mania for conquest (or glory).

These basic causes have continued to remain valid in one shape or another. During the same period of history, Carl von Clausewitz put forward the thesis of "war as an instrument of policy", of war as "political discourse" by other means.

It also implied that the waging of war was to achieve certain ends at the enemy's expense or to prevent him from achieving his. The greatest rationalization and motivation of war, therefore, was to emerge in the Clausewitzian paradigm through the promotion of national interests, both tangible as well as intangible. Ethnocentrism and self-image (including sovereignty, territoriality and national strength), as well as national ideologies, go far to explain the interpretation and perceptions of national interests. There is inevitable scope for misperceptions and miscalculations or even errors of judgement leading to initiation of armed conflicts. Wars may also be initiated accidentally, although the scope for accidental wars would appear to be reducing with modern communication and information systems.

These, and many more deep-rooted complex causes, may be identified as being responsible for providing the motivation for precipitating inter-state conflicts. The cause(s) of war, however, requires a driving force to translate the cause into effect. To that end, availability of military power becomes a pre-requisite. Availability of military power or capabilities make it possible to pursue intentions arising out of one or more sets of causative factors for armed conflicts. Military capabilities open up options; without such capabilities there may be few options. Nations, therefore, constantly strive for ever higher levels of military capability, the process acquiring a momentum of its own. This momentum is further reinforced by threat perceptions and uncertainties about future scenarios, based on elements of distrust and insecurity intrinsic to the very concept of sovereign nation states.

A combination of distrust and perceived threats (increasing the sense of insecurity) and military power (making available response options) can lead to pre-emptive or preventive wars. It would appear that the German attack on Belgium and France in 1914, which came as a response to Russian mobilization, was based on the assumption that Great Britain, France and Russia were about to initiate hostilities. Nations may resort to war if they perceive, rightly or wrongly, that war is inevitable. There is also a correlation between vulnerability and preemption or surprise attack. Vulnerability and an increasing sense of insecurity increase pressures towards greater defensive preparations. Such preparations are normally undertaken with the explicitly-stated objective of avoidance of war through deterrence. However, most cases of improving military preparedness in themselves can increase the insecurity of other states and the probability of war. Nations rarely seem to believe that a potential adversary may be merely seeking security for itself by improving military capabilities as a reactive measure. Lack of transparency in the adversary's military efforts and capabilities, and the absence of trust in his intentions, aggravate insecurities, thereby providing incentives to improve one's own capabilities for possible responses in uncertain – and worst-case – scenarios. An action-reaction phenomenon can then emerge very rapidly in the form of an intensified arms race.

It would thus appear that nations seek military capabilities for what they perceive to be their own legitimate security interests, aims and objectives. This tends to set off an action-reaction process in most cases. Acquisition of arms plays an important part in creating the military capabilities, and all countries seek to keep improving their military capabilities, to which end procurement of arms constitutes a key factor and function. Most of the Third World perforce does so through import of arms. However, the important aspect to note from the point of view of our present analysis is that arms (and military capabilities which they help to create) by themselves do not become a *cause* of armed conflict, except in a very small number of cases. And even in those cases it is the perceptions (rather than realities) of the adversary's military capabilities and distrust of his intentions which contribute to the risk of inter-state conflicts. Wars really start in the minds of men; and arms essentially only provide the means. Wars, therefore, can be prevented only by change in the minds of men: their attitudes, perceptions and belief systems, including those affecting the use and usability of arms.

5. Regional Conflicts

An examination of some of the significant regional conflicts would indicate that in very few cases can the arms trade be considered a factor in the initiation of conflict (see Table 1). These cases are discussed below.

The conflict in Afghanistan is now nearly 15 years old. It assumed an inter-state character with the participation of external powers – the Soviet Union on one side, and the United States, Pakistan and Saudi Arabia on the other, with varying forms and methods. The conflict was really aggravated after the Soviet military intervention in December 1979. This led to the Afghan conflict becoming a symbol and to some extent the substance of superpower competition and the cold war. Massive quantities of arms have undoubtedly been pumped into the conflict during the 1980s. And certainly the conflict has been sustained by the supply of arms, the superpowers being unable to agree on curtailing arms supplies, in spite of agreeing to underwrite the Geneva Accord of April 1988, which sought the de-escalation and termination of the conflict. However given the dynamics of the conflict, it is difficult to see how the conflict might have been *prevented* by controlling the arms supplies.

Similarly, the arms trade can hardly be perceived to have played any role in the *initiation* of armed conflicts in Vietnam, between China and Vietnam, between India and Pakistan (except in 1965), in Cambodia, in Angola, in Yemen, and between Somalia and Ethiopia, although the arms trade helped to sustain most of these wars. On the other hand, the cases of the Iran-Iraq war and the Indo-Pakistan war in 1965 are instructive.

Table 1. The Arms Trade Factor in Some Significant Regional Conflicts (1964–1988)

	Significant arms Supplies		Arms Trade as a factor in Conflict		
Inter-state Conflicts	Before Conflict	During Conflict	Initi- ation	Sustain- ing	Termi- nation
Indo-Pakistan					
1965	Yes	No	Yes	No	Yes
1971	No	No	No	No	No
Vietnam					
1964–73	Yes	Yes	No	Yes	No
Sino-Soviet					
1969	No	No	No	No	No
Arab-Israel					
1967	Yes	Yes	No	Yes	No
1969–71	Yes	Yes	Yes	Yes	No
1973	Yes	Yes	No	Yes	No
1982	Yes	Yes	No	Yes	No
Sino-Vietnam					
1979	No	No	No	No	No
Falklands					
1982	No	No	No	Yes	No
Iran-Iraq					
1980–88	Yes	Yes	Yes	Yes	Yes
Somalia-Ethiopia					
1977–78	Yes	Yes	No	Yes	No
Yemen					
1979	Yes	Yes	No	Yes	No
Afghanistan					
1973–	No	Yes	No	Yes	–
Angola					
1975–	No	Yes	No	Yes	–
Cambodia					
1978–	No	Yes	No	Yes	–

Pakistan started a massive military build-up after it entered into military alliances with the United States and other countries in 1954. India responded with its own military expansion. However, the 1962 Sino-Indian war created a situation of perceived asymmetries. The Indian military, after its reverses, was in the process of expansion although with little hope of receiving any significant arms supplies even by mid-1964. And it was forced to deploy a significant proportion of its forces on the Sino-Indian border. Pakistan, on the other hand, had consolidated its expansion and had absorbed state-of-the-art arms. The military balance which stood at 2 to 1 between India and Pakistan at the time of independence had eroded to around 1.5 to 1, with qualitative superiority of Pakistan in tanks (the M-47 Patton), supersonic fighters (the F-104) and submarines. There were many structural factors which contributed to the war. But the arms build-up, and the perceived favourable asymmetry as a consequence of it, were substantive incentives for Pakistan initiating the war. The arms embargo also, as a consequence, contributed to conflict termination, especially after Pakistan lost the armoured battle in which a major portion of its main battle tanks were destroyed or captured by India. It is also notable that significant arms imports have taken place in Pakistan and India during the 1980s. The structural factors responsible for earlier conflicts have not disappeared; but asymmetry has been absent from the situational factors. This explains why regional stability is now in its 18th year – the longest unbroken period of peace (even if uneasy at times) since independence.

The arms trade to the Persian Gulf/South-West Asian region during the past 15 years is also instructive. In many ways, the arms race in South-West Asia started with the petro-dollar boom of the early 1970s in general, and with the US strategy of building up Iran to play the role of the "regional policeman" in particular. Starting in 1974, Iran under the Shah heavily overarmed itself when it imported arms worth $10.5 billion in the ensuing six years. The arms were acquired mostly from the United States which even departed from its normal policy by providing state-of-the-art, force structure equipment, such as the leading fighter, the F-14 Tomcat equipped with beyond visual-range (160 km) Phoenix air-to-air missile – a weapon system not supplied even to NATO members. Iraq reacted with a similar build-up two years later when it inaugurated a six-year programme involving the import of arms worth $10.4 billion, mostly from the Soviet Union. The superpower cold war linkages with regional disputes resulted in heavy militarization and eventual armed conflict in the shape of the longest war in this century. The Iranian revolution, military purges, and the US embargoes severely curtailed Iranian military capabilities. The earlier arms build-up now became an additional source of vulnerability. (For example, the 77 US-supplied sophisticated F-14 Tomcat fighters worth $34 million each, were unable to use the Phoenix missile because the Americans walked away with the computer interface). Iraq perceived a favourable asymmetry which encouraged it to initiate the war. The war itself was sustained for eight years by arms supplies from diverse sources in spite of the arms embargo placed on Iran. The "loss" of Iran as a surrogate, and Soviet military intervention in Afghanistan, then prompted the United States to build up Saudi Arabia (from 1978) on one side, and Pakistan (from 1981) on the other, besides establishing the Central Command in January 1983 to plan and conduct operations with the Rapid Deployment Force, which by 1989 would consist of 440,000 personnel. Between 1974 and 1986, these three countries alone imported weapons worth $95 billion. The South-West Asian arms race also provided an opportunity for China to enter the arms export market in a major way, with far-reaching consequences. By 1986, China was believed to be exporting arms worth 2 billion dollars per year. In 1988 it supplied CSS-2 Intermediate range Ballistic Missiles to Saudi Arabia who also signed a $30 billion arms import deal with the West. The oil revenues were thus frittered away in a futile arms race and a long, expensive war. The staggered commencement of the military build-ups, the scale (and sources) of arms imports, and the conflict/tensions in the region, indicate the real dimensions and implications of the conventional arms race. It is also significant that with the sharp drop in oil prices, the heavy strains on the economies of these three countries forced them not only to reduce arms imports significantly after 1987, but also helped bring about a cessation of what otherwise appeared to have become an interminable war. Political initiatives by, and a convergence of interests of, the two superpowers also played an important part in not only bringing this conflict to a close, but also in easing tensions all round. Incidentally, the intense arms race in South-West Asia and the geo-political imperatives in the region also triggered off another (though much lesser) reaction. The military build-up of Pakistan, especially with sophisticated weaponry, by the United States after 1981 led to a reaction by India, starting in 1982, also to go in for more modern weapon systems; except that in this case the modernizations and re-equipment programmes held over from the 1970s were telescoped into this new imperative of a degraded security environment. Inevitably, tensions mounted between India and Pakistan.

At another level, the crisis and conflict within Cambodia, and the Vietnamese military intervention there, triggered off not only the Chinese attack on Vietnam in 1979 to "teach it a lesson", but also caused a mini-arms race in South-East Asia. Commencing around 1980, the countries of the Association of South-East Asian Nations (ASEAN) set about military expansion and build-up (most of the countries doubling them in the five years between 1980 and 1985), culminating in Malaysia's £1.8 billion arms deal with Great Britain and in the Australian defence posture shifting to the north and to the north-west after 1986.

6. Controlling the Arms Trade for Conflict Prevention

The dynamics of arms production, and procurement (including imports) are energized by a large number of factors – economic, technological, bureaucratic and political. A detailed assessment of this aspect will not be attempted here. However, two issues merit special note with respect to whether the arms trade can be controlled. First, there is the technological imperative. Advances in technology keep creating asymmetries in relative military capabilities in systemic groups of nation states. This in turn influences the second imperative – that of security.

The technological imperative requires replacement of equipment because of obsolescence which creates vulnerabilities through adverse asymmetry. However, at the root of the technological imperative is the thrust for higher levels of performance and effectiveness; it acquires a momentum of its own, resulting from extensive research and development (R & D) in the military fields. The developed countries, which account for 94 per cent of the world's R & D expenditure, are pursuing continued technological progress in civil and military fields. Nearly 40 per cent of the R & D expenditure is military-related. The US

Commission on Integrated Long-term Strategy, in its report entitled *Discriminate Deterrence* which advocates a future strategy for the United States, strongly supports a further impetus to military R & D so that "In the Third World, no less than in developed countries, US strategy should seek to maximize our technical advantages" [1] and futuristic technologies are pursued. In 1980 world military expenditure was of the order of $78 billion; and the figures have been rising ever since. (In comparison, the developing countries, encompassing 75 per cent of the world's population, spent less than $5 billion on military-related R & D. With such levels of investments in military R & D, arms production of increasingly sophisticated weapons and their transfer to developing countries are likely to remain a significant reality.

Developed countries also employ arms transfers as an instrument of policy, besides seeking economic advantage. For example, the principles enunciated in *Discriminate Deterrence* include the recommendation for the United States: "Our means should include:

- Security assistance at a higher level and with fewer legislative restrictions that inhibit its effectiveness.
- Versatile, mobile forces, minimally dependent on overseas bases, that can deliver precisely controlled strikes against distant military targets.
- Allies that help us defend common interests beyond alliance boundaries.
- In special cases, US assistance to anti-Communist insurgents who are resisting a hostile regime imposed from the outside or a regime that threatens its neighbours. The Free World will not remain free if its options are only to stand still or retreat."

The recipient countries have their perceived needs and justifications. Arms procurement by the developing countries, however, cannot be treated on a separate rationale from that by the developed countries themselves. In fact, most developing countries are generally responsive to strategic doctrines and generation of arms in the developed world. In particular this is true of the militarily-more-significant powers. And unless the basic issue of the causes of insecurity and vulnerability in the developing world is adequately addressed, it would be overly optimistic to expect that the arms trade, which is a very minor aspect of the overall process of generation and stockpiling of arms, can be controlled.

The only area where some curbs can perhaps be placed on the arms trade so that conflict prevention receives a positive contribution, is to control trade in offensive weapons. Offensive capabilities have greater potential for increasing the sense of insecurity of other states, thereby forcing them to find suitable responses. Curbs on them would, therefore, enhance the security of all countries. It is not easy to find adequate criteria for the classification of weapons as offensive, but this is not impossible. Efforts are required to establish scientifically criteria for offensive capabilities and to negotiate suitable control and verification mechanisms. Such controls, of necessity, will also need to be global in nature; and the type of control most likely to have a durable effect would have to cover development and production of such weapons.

Experience of inter-state conflicts during the past 25 years indicates that the arms trade *per se* has been a marginal factor in the outbreak of conflicts, although it has helped to sustain most conflicts.

It would thus appear that controlling the arms trade is likely to make a very limited contribution to the prevention of conflicts. In the three identifiable instances where the arms trade did contribute to the initiation of war, certain situational factors were present. The most important of these was perception of favourable asymmetry on the basis of arms received. However, the situational factors could operate only because of the existence of structural factors as the causes of conflict. Even in the situational factors, arms supplies were but one of many factors, although they provided the means to translate cause into effect. On the other hand, it can be argued that in many cases the acquisition of arms in conditions of greater symmetry has helped to maintain systemic stability and, therefore, has contributed to the prevention of conflicts. This would suggest that controlling the arms trade so as to enhance stability though symmetry can make a contribution to conflict prevention, provided other situational and structural factors also operate to attenuate conflict potential.

It is, however, in the field of intra-state conflicts where the complexities of the arms trade and their impact become extremely difficult to unravel. This is especially so because of the socio-political ramifications of these conflicts, on the one hand, and because of the existence of "grey" and "black" markets in arms, in addition to the official arms trade, on the other. The accessibility of sophisticated arms to terrorist groups has added a new dimension to the problems of peace and security, not only within the state but spilling over to other states and regions. Vigorous measures for the control of all forms of the arms trade, especially in sophisticated weapons, are thus required to prevent intra-state conflicts. To achieve any modicum of success international control mechanisms are necessary at the global level.

References

1 "Discriminate Deterrence" *Report of the Commission on Integrated Long-Term Strategy*. F. C. Ikle & A. Wohlstetter, January, 1988. (Washington, DC: US Government Printing Office)

Part VI

Role of United Nations and Multilateral Organizations in International Security

The Role of the United Nations and Multilateral Organizations in International Security

Omran El-Shafei

In a report published by the News Agency *Reuters* on major events during the year 1988, the United Nations (UN) was referred to as the World Organization which resumed its central diplomatic role after years of marginal impact. The achievements of the UN over the last couple of years were also recognized by the award of the Nobel Peace Prize for its peace-keeping operations. The report emphasized the UN's successes in bringing closer to peace the situations in Afghanistan, the Gulf, Cyprus, Southern Africa and the Western Sahara. These successes could not have taken place except for, and as a result of, improved relationship between the two superpowers. This improvement also led to a change of attitude towards the UN and its global role. The report concluded by stressing the point that in spite of these successes, the UN still finds itself in a precarious financial situation, particularly if it has to undertake the more ambitious and costly peace-keeping operation in Namibia.

This paper tries to deal with the impact of world developments on the desired instrumental changes in the International Organizations and particularly the UN, the impediments and hindrances that work against a more effective UN, and the need for a common approach on the part of the international community in matters of security and cooperation.

Because the United Nations came to existence after a long and ferocious world war, in which most of the nations existing at that time were actively involved or affected, and in which millions of people lost their lives, the World Organization reflected the realities of that time. In particular, there was recognition of the need to restructure relations between the nations on a more sustainable basis, taking into consideration the lessons drawn from the previous experience of the League of Nations. This led to the drive for universality and the establishment of principles of equality among members of the World Organization; to acceptance of the principle of non-intervention in internal affairs; and to the entrenchment of the principle and practice of international cooperation as a corner-stone in the relations between nations.

It is obvious that the United Nations we are talking about is no longer the same. Not only has its membership tripled but, what is more significant, its work, together with that of the specialized agencies, has become so varied and comprehensive that very few human activities can be excluded from the UN's sphere of interest and concern. Moreover, the UN not only extended its active involvement in World affairs, but it had to face continuous and rapid changes and developments at the national, regional and international level, and adapt itself to these challenges. Even many critics of the UN and the whole system of International Organizations agree that in spite of its deficiencies, the UN has continued to be a valuable public forum for all its members to air their grievances and to mobilize World Public Opinion in support of its views.

It may be relevant, at this juncture, to refer briefly to the major criticisms levelled against the UN as reported by Western media:

- the inability of the UN to act swiftly and firmly, particularly at times of crisis involving armed conflict, the reasons being always major differences between the two superpowers;
- the inability of the UN to implement its decisions, particularly those relating to sanctions and other punitive measures;
- a great part of its regular budget is spent on the running of the administration and very little goes to meaningful projects;
- the Third World member countries enjoy a disproportionate influence, not commensurate with their responsibilities within and outside the UN, and that this numerical strength is always used detrimentally to the interests of certain major powers in the Organization.

These criticisms and other similar ones should not be the issue, for the single reason that there always will be the feeling that the UN has not lived up to the expectations of one group or another of its members, due to the varied and complex nature of the interests of these members. However, such criticisms could yield positive results if they were seriously debated and dealt with in an objective and democratic fashion.

Admittedly, there are many factors which have contributed to the weakening of the UN, or at least to the lessening of its effectiveness. Most observers agree that the East-West conflict, and, in particular, the differences between the two superpowers have played a major role. Then there is the North-South dilemma as an area of discord and disenchantment, particularly on the part of the developing countries. The external debt problem of the developing countries has burdened the entire system, and could – if it continues in the absence of a global programme for the restructuring of the world economy – threaten the existence of the whole international system.

Many observers believe also that the process of decision-making in the UN, which can be long and frustrating, deprives the Organization of the opportunity to respond immediately to major world events. Lack of respect for UN decisions adds to the frustration of World Public Opinion which tends to underrate the Organization, particularly at times of crisis. The frequent use of the veto on the part of certain members of the Security Council could be taken as a sign of disrespect for the whole system. Sometimes, the whole system is also blamed for the meagre results when it comes to lengthy and extraneous negotiations dealing with arms control or with economic adjustments. This type of criticism has to be examined very carefully for the issues involved are quite intricate and complex.

Another element weakening the UN is the tendency to bypass it. This is particularly the case with respect to the Security Council when it comes to issues involving settlements of disputes. Regional organizations may have a better chance of resolving some of the regional problems if submitted to them. But the question remains whether it is possible to ignore the UN and the channels it offers.

A final point of importance concerns the policy of denying the whole system of financial support, by the tactic of delaying payment of membership or voluntary contributions, as a means of exerting pressure. This policy has always proved to be counter-productive and very rarely achieved its purpose.

Despite these obstacles, sometimes insurmountable, the UN has survived the last four decades. Obviously it could not have done so had its members found it not in their interest to keep it going, whether in playing a central or a peripheral role. There are indeed good reasons for keeping the UN and enhancing its capacity. First, there is the increasing awareness and conviction among its members, particularly the superpowers and their allies, of their common interest in preserving the present international system. This conviction becomes more entrenched when these powers face real dangers. The danger of the nuclear weapons to mankind and to the survival of human civilization has been recognized not only as such, but it also led to the adoption of policies that address the common interests of nations rather than their individual and immediate self-interest. A second reason for strengthening the UN is that most of the current malaise has its roots in social and economic tensions. The question is whether the UN system as a whole is equipped to deal with these problems. As stated earlier, many achievements obtained through the present multilateral organizations have been valuable. One has only to refer to a few examples: codification of principles of international law; laying down common international standards and instruments for the protection of human rights; and cooperation in world communication systems. On the other hand, real and genuine achievements, in such matters as increasing international trade through the elimination of barriers, promoting world economic and social development particularly for the benefit of the developing countries, and reforming the world monetary system, are still lagging behind.

Many observers believe that the potentialities of the system have not been fully realized. They even argue that as long as major issues of security, arms control, human rights and economic and social development are dealt with separately, there will be no attempt to attack disorders at their roots. It is true that the system has been working with consultation taking place among the separate and independent components. From time to time, joint meetings between the Directors-General of various entities try to coordinate and harmonize their work, but it has never been suggested that such an exercise could bring about a global plan for adjustment and reform.

What is needed is a change in the conceptual attitudes of the member states towards the functioning of the whole system. It is not enough for the world community to envisage efforts for the maintenance of world security through resolutions of the Security Council and the utilization of UN peace-keeping forces, or through resort to the good offices of the Secretary- General. The content and conception of security policies have to reflect the changes in the conception of economic relations among states. Fundamental changes are taking place in international relations, and these changes should lead to more confidence and trust among nations; the multilateral organizations, such as the UN, can offer the modalities for coping with such changes. States should continue to use the facilities presented by the entire system.

Proposals for the reinforcement of the capacities of the multilateral organizations should be considered seriously. The suggestion to create a War Risk Reduction Centre, attached to the office of the UN Secretary-General, for the purpose of making suggestions to the Security Council, is an idea that merits attention. So is the proposal to revitalize the Military Staff Committee (Article 47 of the Charter). In every domain of the UN and its specialized agencies, there are similar proposals to enhance their roles; such proposals should be considered more thoroughly with a view to their eventual implementation.

As the Secretary-General of the UN stated in his introduction to the report on the work of the Organization during the year 1989:

> "*Changes in perceptions and attitudes suggest that we may be witnessing a transition, however slow or occasionally uncertain, towards a new pattern of relationships at the global level. The transition has the logic of necessity behind it. It is certainly justified by the insupportable cost and the incalculable dangers of a self-propelling arms race. It could derive support from the realization that security cannot be viewed in military terms alone nor does the application of military power resolve situations in traditionally expected ways. It is or can be propelled by the need of greater attention to the problems of economic modernization, to the social problems that economic growth has left untouched. All these and other factors combined with the technological revolution and the sense of global interdependence seem to call for radical adjustments of outlook on the part of the World's leadership. There are, of course, no guarantees against temporary reverses or setbacks in the process, nor can ambivalence in the relationships of power blocs be excluded. However, the direction appears to be better set and helped by weightier factors more now than at any other time in recent years. How this transition will affect the UN are questions of practical importance which merit the most serious reflection on our part.*"

Regional Organizations and Their Contribution to International Security: The Case of Asean

Muthiah Alagappa

1. Introduction

The Association of Southeast Asian Nations (ASEAN) is not a conventional security organization, but since inception its core goal has been the promotion of peace and stability in Southeast Asia. Almost every declaration of the Association has made explicit mention of this goal. In enumerating the aims and purposes of the Association, the founding Bangkok Declaration of 1967 accords primacy to this objective; the Kuala Lumpur Declaration of 1971 is exclusively devoted to the management of political and security relations with external powers and to the role of regional cooperation in promoting "lasting peace in Southeast Asia"; and the only regional treaty sponsored by the Association is on the settlement of political disputes among countries in the region. A very substantial part of the Association's efforts in the last decade has been engaged by the conflict in Cambodia.

This paper explores ASEAN's contribution to peace and stability in the first place in the ASEAN region, and secondly in Southeast Asia. This distinction is necessary because, despite the declared intention to encompass all states in the region, the Association continues to be a sub-regional organization comprising the non-communist states in Southeast Asia apart from Burma. The paper will also explore the possibilities and constraints in enhancing ASEAN's contribution to security in the region in the light of the changes taking place in major power relations, as well as the emerging distinct possibility for a settlement in Cambodia.

The concept of international security draws attention to the anarchic structure of the international system, the competition and conflict it fosters, and the endemic fear for and consequent concern to preserve and enhance national security. In a condition of anarchy, war can occur with force becoming the *ultima ratio* in international politics. Unlike the selfhelp prescriptions of the realist school, which focus on the power of states, international security strategies and policies are concerned with controlling and minimizing the negative consequences of international anarchy, especially the war-proneness of the system. The goal is not transformation of the structure of the system, but through the institution of international rules, norms, and procedures, to make interstate behaviour more orderly and predictable, and reduce the role of force. This is precisely the goal of the United Nations at the global level. The UN Charter envisages a role for regional organizations in promoting international security. It encourages pacific settlement of disputes at the regional level, but subjects regional enforcement action to the sanction of the UN Security Council.

In promoting peace and security, regional organizations may perform the following functions:

Civil Order. They can seek to prevent the outbreak of disputes, prevent degeneration of disputes into conflicts, and provide mechanisms for resolution of existing disputes and conflicts among member countries. Domestic conflicts in member countries should technically be excluded from this function. In fact, one of the fundamental principles to be upheld by regional organizations in promoting regional security should be non-interference in the internal affairs of member states. However, the possible negative consequences of domestic conflicts for regional order and security imply that the regional civil order function cannot be neatly separated from domestic civil order in member states. A delicate balance has to be drawn between "negative" and "positive" interference. But this is extremely difficult, if not impossible, and would pose a dilemma for regional organizations, specially in the developing world where states are not only weak but also "nationalistic".

External Security. They can seek to provide for the security of member states from external threats. This could be achieved by eliminating or minimizing sources of threat, by providing mechanisms for peaceful resolution of disputes and conflicts that have emerged, and in the event such mechanisms are inadequate, by harnessing the necessary power to defend a member state that is confronted with external threats to its security.

Welfare Function. They can seek to promote the socio-economic well-being of member states through intraregional economic cooperation which would also assist in reducing the exposure of member countries' economies to the vagaries of the international system over which they have little or no control. This would enhance domestic stability and facilitate the development of national resilience in member states.

ASEAN's contribution to the security of its member states may be evaluated through a discussion of the progress made in relation to these three functions. Its contribution to peace and stability in Southeast Asia as a whole may be explored through a discussion of the impact it has had on the strategic environment in the region, especially on the pattern of cooperation and conflict, and on the behaviour of external actors in the region.

2. Contribution to Security in the ASEAN Region

2.1 Civil Order Function

ASEAN has made a substantial contribution to improvement in bilateral relations amongst its member states. In the mid-1960s, relations involving four of its five (now six) members were characterized by much tension and conflict. Specific mention must be made of the poor relations that existed between Malaysia and Indonesia, Malaysia and Singapore, Malaysia and the Philippines, and Singapore and Indonesia. Except for the Malaysia-Philippines relationship, all others have improved quite substantially. Although it is difficult to attribute the improvements that have taken place entirely to ASEAN, it is also difficult to deny that the Association has made a significant difference. As observed by many ASEAN leaders, the Association, especially the elite socialization that it facilitated, has been an important factor in lubricating relationships that could otherwise have generated frictions.

The relations between Malaysia and the Philippines continue to be less than cordial. The contentious issues are the Philippine claim to Sabah, the tens of thousands of illegal Filipino immigrants in Sabah, the perception in Manila that certain quarters in Malaysia support the Muslim secessionist movement in the southern Philippines, the conflicting claims in the South China Sea, piracy and smuggling.

The Philippine claim to Sabah appears to be symbolically the most important, having been the cause of a break in diplomatic relations on two occasions (1963–66 and 1968–69), and threatened the survival of ASEAN. Ferdinand Marcos in 1977 stated that in the interest of ASEAN solidarity and harmony, the Philippines would take necessary steps to drop the claim to Sabah. But he did not follow through with the required legislative action. Bilateral discussions in 1986–87 resulted in an understanding to the effect that following Manila's renunciation of the claim, Malaysia would sign a treaty of friendship and economic cooperation and an agreement on joint border security cooperation with the Philippines. President Corazon Aquino, rather hurriedly before the third ASEAN summit, submitted a bill renouncing the claim which passed the House of Representatives but not the Senate. With the ASEAN summit out of the way the urgency to resolve the dispute appears to have diminished. The Philippine Senate in May 1989 rejected the Aquino bill stating that Kuala Lumpur must first settle the proprietary claims of the heirs of the Sultan of Sulu, (who "leased" the territory to the British traders in the 19th century), sign a treaty of friendship and cooperation, and forge an agreement on joint border patrols, before Manila can renounce its claim. Malaysia does not want to link the three issues. Hence the prospects for an early settlement of the dispute are not bright.

Meanwhile, relations between Malaysia and the Philippines took a turn for the worse in 1988 in the light of claims and counter-claims concerning piracy and military action by each other, and the erroneous claim by some Philippine political leaders and officials that Malaysia had illegally occupied some islets belonging to the Philippines. It was subsequently discovered that the Philippine leaders had used an inaccurate (navigational) map and an official statement was issued to retract earlier Philippine accusations. Coming at the times of the second and third ASEAN summits respectively, sensitivity to the group dynamics of ASEAN would appear to have been a factor in the Marcos announcement about dropping the claim to Sabah, and in Aquino's attempt to resolve the dispute, but domestic considerations in the Philippines appear to be working against a final resolution. However, unlike in 1963 and 1968, events in 1988 were not allowed to get out of control. The Philippines is not actively pursuing its claim but is having political difficulty in dropping it. The visit of Mahathir Mohamad to Manila to attend the Third ASEAN summit was a positive development. Although this is not the same as a bilateral visit, it reflects progress in the light of the fact that earlier prime ministers had avoided visiting Manila. Relations between the two countries are, however, not cordial and the Sabah dispute remains contentious with potential to disrupt ASEAN.

Another troublesome relationship in which considerably more progress has been made, and in which ASEAN has been a significant factor, is that involving Malaysia and Singapore. Membership of ASEAN has progressively contributed to the political and psychological security of Singapore, and this in turn has contributed to better relations between Singapore and Malaysia. In the absence of formal bilateral agreements, ASEAN has also provided the framework for cooperation and accommodation and for third party mediation during periods of crisis. Elite socialization, especially the close rapport between Lee Kuan Yew and Mahathir, has been very important in the positive upturn in relations between the two countries during the 1980s. This relationship, however, is not completely free of tension, as reflected in the "crisis" that developed following the adverse reaction in Malaysia (and to a lesser degree in Indonesia) to the visit of the Israeli president to Singapore, and to comments by Goh Chok Tong and Lee Hsien Loong on the position and role of the Malay community in Singapore. Given the structural character of the issues that can make for tension between these two countries, it is likely that the relationship will continue to be characterized by some mistrust and suspicion. But the important point to note is that the "crises" were controlled and resulted in extensive bilateral consultations to improve the foundations of bilateral relations. As a result of these consultations, bilateral relations have not only returned to normal but have in fact improved as demonstrated by the recent signing of the gas and water agreements and the initiation of bilateral military exercises; this would have been unthinkable even a few years ago.

While ASEAN has played an important role in preventing the degeneration of bilateral tensions and disputes into conflicts, its role in resolving outstanding disputes has been limited. The 1976 Treaty of Amity and Cooperation makes available a mechanism for pacific settlement of disputes among signatory countries, and the ASEAN Task Force Report, which was endorsed by the seventeenth ASEAN Ministerial Meeting, requests member countries to 'bear in mind the value of the machinery relating to the pacific settlement of disputes among signatory countries, and the ASEAN Task Force Report, which was endorsed by the seventeenth ASEAN Ministerial Meeting, requests member countries to "bear in mind the value of the machinery relating to the pacific settlement of disputes as provided in the Treaty ..." but the provisions remain unutilized. Thai and Indonesian efforts at mediation in the Sabah dispute have been partially successful but have not resulted in full resolution and, as pointed out earlier, Malaysia and the Philippines have not been able to resolve the issue on a bilateral basis. ASEAN has also not been able to prevent the emergence of new disputes amongst its member states. For example, its members took unilateral action in declaring their exclusive economic zones resulting in mutual protests and conflicting claims. Nor has ASEAN been a force in the search for a settlement of these conflicting claims, which is being pursued essentially through individual country efforts. Almost every ASEAN country is upgrading its military capability to defend its exclusive economic zone and its claims therein. A coordinated approach by its members and agreement to resolve conflicting claims, using the machinery provided for in the 1976 Treaty, would have elevated the stature of ASEAN and also strengthened the position of its members *vis-á-vis* other non-ASEAN claimants.

ASEAN has also been a moderating force in minimizing – if not preventing – interference by members in each others domestic affairs. For example, Malaysia and Indonesia could have chosen to support the Malay-Muslim insurgencies in southern Thailand and southern Philippines. Instead, in the spirit of ASEAN cooperation both these countries counselled respect for territorial integrity and for moderation in the Organization of Islamic Conference (OIC). The Malaysian and Indonesian representation in the OIC has been valued by Thailand and the Philippines. The deteriorating situation in the Philippines in 1985–86, however, presented ASEAN with a dilemma. Applying the principle of non-interference rather rigidly, the other ASEAN countries avoided making any collective public comment on behalf of ASEAN, and eventually in February 1986 issued a mild statement expressing concern and calling upon all Philippine leaders to join efforts to pave the way for a peaceful solution to the crisis. While this may have been technically correct, it highlighted the complicated overlap of domestic and regional civil order functions and revealed the limits of ASEAN in maintaining civil order under such conditions. The fragility of domestic consensus and regime legitimacy in a number of ASEAN countries, and also the lack of any real capability to influence the situation in the Philippines, prevented more specific action. ASEAN was virtually reduced to the status of a bystander in a development that could have had consequences for the Association had the turn of events in the Philippines taken different directions.

In sum, there is little doubt that ASEAN has contributed to the creation of goodwill, trust and confidence amongst its members, although its importance *vis-á-vis* other contributory factors may be difficult to identify and "calculate";

it facilitated the transformation of a region of "turmoil" into a more predictable and stable one in which the role of force has been minimized but not eliminated. It may be pertinent at this juncture to make a brief comment on the increasing reference to ASEAN as a "partial or semi-pluralistic security community". While this may not be wholly inaccurate, there are reasons for exercising caution in embracing such an optimistic characterization. Much of the improvement that has occurred in relations among the ASEAN countries has been a function of elite socialization, and does not reflect the growth of a more broadly-based sense of community. Given the relatively low level of political institutionalization, and the lack of experience in peaceful political succession in a number of ASEAN countries, the emergence of another Sukarno or Marcos, and the consequent negative impact of ASEAN cooperation, cannot be precluded. It should also be noted that the defence purchases and plans of a number of countries cannot but be explained with some reference to the threats they continue to perceive from their fellow ASEAN neighbours. The intention here is not to negate the earlier positive assessment of ASEAN's contribution but merely to underscore the qualifications that must accompany characterizations such as a "partial pluralistic security community".

3. External Defence Function

The external defence function was implicit from the very beginning of the Association. Indonesia had wanted to accord greater emphasis to this function but the other countries were less enthusiastic. The Bangkok Declaration thus makes only marginal mention of it. ASEAN members continued to value their pre-ASEAN security connections. Several developments including the UK decision in 1969 to accelerate the withdrawal of British troops from east of Suez, the Guam doctrine, Sino-American rapprochement, and the American withdrawal from Indochina, undermined the faith of ASEAN countries in their traditional defence links and underscored the need for greater self reliance. ASEAN became more concerned with the international environment, and the ASEAN foreign ministers agreed in 1972 to meet in special sessions to review international developments affecting the region.

ASEAN did not take the route of a multilateral security pact. From the beginning, ASEAN, for several reasons, was keen to avoid projecting the image of a military-security organization, although its detractors (the Soviet Union, China and Vietnam) did perceive it as such. Despite repeated calls from within and without, the Association has deliberately refrained from instituting an ASEAN security arrangement encompassing all its members, although since 1976 it has endorsed security cooperation amongst its members on a non-ASEAN basis.

ASEAN's approach to the external security function has emphasized the development of an internationally-recognized peace zone in Southeast Asia. In

1971, the ASEAN members declared their intention to secure recognition of Southeast Asia as a zone of peace, freedom and neutrality (ZOPFAN), free from any form of interference by outside powers. Implementation of this idea is deemed to require the elimination of foreign military presence in Southeast Asia, termination of security arrangements with external actors, creation of a nuclear-weapon-free zone, and the institution of an internationally-accepted mechanism for dispute and conflict resolution. ASEAN began to develop a specific proposal for a nuclear-weapon-free zone in 1985, but no specific programmes have been drawn up for the other elements. Because of lack of agreement on the specifics among ASEAN members, issuing primarily from uncertainty as to whether the concept can really assist in the preservation of their national security, because of non-support from the other Southeast Asian states due to the Cambodian conflict, and because of opposition from at least a couple of external powers, ZOPFAN has continued to remain on the backburner.

Even when presented with the opportunity, ASEAN has been unable to formulate common policies on issues that have a bearing on its goal of ZOPFAN. For example, on the development of relations with China, a country which was and continues to be viewed by almost all the ASEAN states in differing degrees as a threat to their national security, a country which would no doubt continue to have a significant bearing on the Southeast Asian strategic environment, the ASEAN countries have not been able to develop a common policy. Until recently (February 1989) Indonesia was not disposed to resume diplomatic relations with China, whereas Malaysia, Thailand and the Philippines had established diplomatic relations in 1974–75. Singapore, for several reasons, stated that it would be the last ASEAN country to establish diplomatic relations with China. Notwithstanding the absence of formal diplomatic relations, Singapore's relations with China are well advanced and perhaps the best in Southeast Asia. Thailand's relations with China are also good and include a substantial military dimension which is of concern to Malaysia and Indonesia. The latter two countries, in the light of domestic and other considerations, view China quite differently, and this is reflected in their cautious approach to relations with that country. Thus, differing national imperatives make harmonization of policy in respect of an important external element rather difficult. This would become even more complicated if and when ASEAN is expanded to include the countries of Indochina. Yet, this is an important prerequisite for the implementation of ZOPFAN.

Similarly, ASEAN has not been able to make use of the opportunities, such as those presented by the upcoming renegotiation of the Military Bases Agreement (MBA) which expires in 1991, and Mikhail Gorbachev's offer to close bases reciprocally, to formulate a common policy giving specific content to the statement in the Bangkok Declaration that all foreign military bases in the region are temporary in nature. The attempt by the Philippine Foreign Minister to obtain ASEAN support for the continuation of the American bases was unsuccessful, although Singapore, Malaysia and Thailand expressed their support individually with differing degrees of enthusiasm. Indonesia stated that it was not opposed to the bases as long as the Philippines deemed them necessary. In other words, it was a matter for the Philippines to decide. While undoubtedly the final decision would be that of the Philippines, ASEAN has failed to make use of an opportunity to steer the decision in a desired direction. It does not appear to be able to muster the necessary political will and capacity to realize ZOPFAN which, as conceptualized by ASEAN thus far, also appears quite inadequate to provide for the external security of member states.

ASEAN has acted with greater urgency and effectiveness in responding to developments in Indochina. In assessing and reviewing the implications of the Agreement on Ending the War and Restoring Peace in Vietnam, the ASEAN foreign ministers meeting in special session in 1973 agreed that, pending the convening of a Southeast Asia forum of all states in the region (to discuss matters of vital interest and common concern), and expanding ASEAN to include all Southeast Asian states at an opportune moment, the ASEAN countries should promote bilateral relations and participate in the rehabilitation and reconstruction of all Indochina states.

In 1975, ASEAN responded to the communist victories in Vietnam, Laos and Cambodia, by demonstrating solidarity and cohesion. The ASEAN Heads of Government met for the first time in a summit meeting and signed the Declaration of ASEAN Concord which called for the exertion of efforts to create a strong ASEAN community that would be respected by all and respecting all nations. For the first time, political cooperation received explicit mention with a relatively specific agenda, security cooperation was endorsed, but with the proviso that it was to be conducted outside the ASEAN framework and the Concord outlined a relatively comprehensive agenda for economic cooperation. The Treaty of Amity and Cooperation, also signed and ratified in 1976, codified the norms for conduct of interstate relations and outlined the procedure for pacific settlement of disputes. The Treaty was left open for accession by other Southeast Asian states. (The forthcoming ASEAN Ministerial Meeting in Brunei is expected to modify Article 18 of the Treaty to provide for accession by non-Southeast-Asian states like Papua New Guinea). These actions were intended to demonstrate ASEAN political resilience and resolve. Concurrently, ASEAN continued its efforts to engage the states in Indochina in constructive relations.

The Vietnamese invasion of Cambodia in December 1987 and continued domination of that country was viewed as a threat to peace and stability in Southeast Asia, with Bangkok viewing it as a direct threat to its national security. Although previous developments in Indochina were viewed with concern, this particular development, and its linkage to the Sino-Vietnamese and Sino-Soviet conflicts, increased the salience of the external defence function. ASEAN deplored armed intervention in Cambodia, affirmed the right of the Cambodian people to self-determination, and called for the immediate withdrawal of all foreign troops. With Thailand in the lead, ASEAN adopted a multi-pronged strategy to make the cost of occupation unbearable, and to force Vietnam to negotiate a settlement based on the principles advanced by the Association. The multipronged strategy entailed denial of diplomatic recognition to the newly-installed Cambodian government, continued recognition of its predecessor, diplomatic and economic isolation of Vietnam, and maintaining military pressure on that country.

On its own, ASEAN did not have the economic and military resources to implement its strategy. It has to harness international support. This was possible because the United States, the countries of the European Economic Community (EEC), Japan and China were in *de facto* alignment against the Soviet Union and Vietnam. Had the international pattern of alignment been different, it is highly unlikely that ASEAN could have achieved the success it did. Despite the divergent viewpoints, especially between Indonesia and Thailand, ASEAN, to its credit, was able to maintain unity of purpose and position throughout, and has by and large succeeded in denying victory to Vietnam.

While the internationalization of the Cambodian conflict was necessary and served ASEAN's immediate purpose quite well, it made conflict resolution that much more difficult, and also reduced ASEAN's role and influence in the resolution of the conflict. ASEAN's strategy was suitable for maintaining the stalemate but not for resolving the conflict. It also rendered the UN forum, which had served ASEAN quite well in keeping Vietnam in the "doghouse", virtually useless in the resolution of the conflict. The peace process, when it started, largely escaped the control of ASEAN. The critical variables have been Sino-Soviet relations, Sino-Vietnamese relations, and relations among the four Cambodian factions and the interests they represent. Improvement in Sino-Soviet relations has contributed to the Vietnamese announcement that all troops will be withdrawn from Cambodia by the end of September 1989. There is as yet no accord on the issue of power-sharing in a post-settlement government in Pnom Penh. ASEAN has been trying to play a role in the ongoing peace process through the Jakarta Informal meeting (JIM) but with limited success. Thailand has also taken its own initiative with some impact because of its key geopolitical position and role in the conflict.

It is pertinent to observe here that from a Thai perspective, ASEAN was a useful politico-diplomatic instrument. The latter's military-security value in the event of a major aggression against Thailand, however, would be marginal, given the meagre security resources commanded by the ASEAN countries, and also because of the uncertainty as to whether these countries would actually provide military support if the need arose. Bangkok looked to the United States and to China, especially the latter, as the main countervailing forces against Vietnam. Sino-Thai military relations began to develop and have now become quite substantial. The other ASEAN countries, less Indonesia, also opted for a higher level of security cooperation with the United States. Singapore and Malaysia reinvigorated security relations with the Five Power Defence Arrangement (FPDA) partners, and Brunei relies on its security arrangement with Great Britain for its external defence. Thus, the role of ASEAN in the external defence function remains limited to the politico-diplomatic dimension. While not unimportant, clearly it is inadequate. The lack of military capability to support ASEAN's diplomatic initiatives has been recognized but there is no consensus on how to proceed. Mochtar Kusumaatmadja, the former Foreign Minister of Indonesia, for example, has suggested the need for ASEAN countries to enter into some form of defence arrangement to strengthen the basis for collective action. Since the beginning of 1989 the Malaysian Foreign Minister Abu Hassan Omar has been suggesting a security accord of some kind among the members of ASEAN in the light of anticipated changes in the region. But it is uncertain whether these suggestions would lead to anything concrete.

ASEAN has been quite successful in formulating and implementing a common policy on the issue of refugees from Indochina. It drew international attention to the issue, to the negative impact on stability of its members states and the region, urged Vietnam to take measures to control the problem at source, and harnessed international support and resources to administer and resettle the refugees who have entered the ASEAN countries. ASEAN played a key role in the various UN-sponsored international conferences on the issue of the Indochina refugees. The outflow of refugees from Indochina, however, has not stopped. A recent count put the number of refugees in camps in Southeast Asia (excluding the Khmer refugees in Thai border camps) at 72 000. The ASEAN countries, perceiving "fatigue" on the part of the Western countries, and to underscore the continuing gravity of the problem, have successfully canvassed support for an international conference in Geneva in July 1989. A preparatory meeting was held in Kuala Lumpur in March 1989 during which the fundamentals of a comprehensive package of action were agreed upon.

4. The Welfare Action

The political and economic benefits of intra-ASEAN economic cooperation have been recognized from its inception and have been reiterated periodically, especially when the ASEAN economies were experiencing slow or negative growth. The Report of the Group of Fourteen on ASEAN Economic Cooperation and Integration (1987) stated:

"*...in the face of structural changes in the world economy which have resulted in a hostile external economic environment that already challenges us all, the economic dimension must be developed to the point where, for the first time, economics will truly be a central game in ASEAN. We believe that if the Manila summit does not grasp the economic imperative, we will have lost a crucial strategic opportunity.*"

With the Manila summit in view many proposals were advanced. Most of them were a variant of the free trade model. The ASEAN Heads of Government, however, on the advice of their economic ministers, opted for a gradual approach. Measures adopted were in the main aimed at improving existing schemes. Whilst they were not insignificant, they fell short of expectations. Many reasons account for the lack of progress in intra-ASEAN economic cooperation, chief among them being the primacy accorded to short-term national interests on the part of countries who fear they would stand to "lose" in opening up their markets: the unacceptability of asymmetrical gains. Intra-ASEAN trade currently only accounts for some 17–19 per cent of total ASEAN trade. Also, only a small number of ASEAN industrial projects have got off the ground.

ASEAN has been more successful in managing external economic relations with its major trading partners. Exploiting its strengths (the dynamic nature of its economies, growing size of its markets and its geopolitical significance) ASEAN has been relatively successful in its dialogue relations with the United States, Canada, the EEC countries, Australia, New Zealand, and especially with Japan, in resolving immediate problems, gaining preferential access to markets, securing greater flow of capital, investment and technology and generally making its concerns heard.

External perceptions of ASEAN as a peaceful community of states making good economic progress have helped to sustain a favourable investment climate. Although it will be difficult to isolate and quantify ASEAN's contribution to a conducive environment, it would be difficult to deny that it is a significant contributory factor.

5. Impact on the Southeast Asian Strategic Environment

Relative success in its civil order function, rapid economic growth enjoyed by its member states, and its ability to act collectively on a number of regional issues, have made ASEAN a stabilizing force in Southeast Asia. This is not insignificant if one takes account of the fact that the Association comprises six of the ten countries in the region and that its members are situated astride the key waterways in Southeast Asia.

ASEAN had become a significant force in the international relations of Southeast Asia which cannot be ignored. Increasingly, American and Japanese Southeast Asia policies are being formulated with ASEAN as an important component. Even China, which was once highly critical of the Association as a "reactionary military alliance directed at China", has come to view it more favourably and supports it as a regional force. Admittedly, it is in the interest of the United States, Japan and China to support ASEAN, this partly accounts for the success achieved by the Association. This, however, does not mean that ASEAN has no leverage, or that it has not been able to modify the behaviour of external elements in its favour.

Without the regionalist credentials of ASEAN and the support of Thailand, China would not have been able to support effectively the Khmer Rouge, and to continue the international isolation of Vietnam for more than a decade. China has had to make some concessions to ASEAN, not only in relation to the Cambodian conflict, but also on other issues, such as its support for communist insurgencies in the ASEAN region; it also had to guard against upsetting the ASEAN countries by too aggressive an assertion of its claims in the South China Sea. Japan has had to modify its Indochina policy, and the United States for its part has become more sensitive to ASEAN concerns about China. Even the Soviet Union, the patron of Vietnam, has conceded the significance and contribution of ASEAN and has sought to enter into a dialogue relationship with the Association. Gorbachev in his message to President Aquino, in her capacity as chairwoman of the third ASEAN summit, indicated the desire of the Soviet Union to become a dialogue partner.

It has been argued that ASEAN's Cambodia policy has contributed to confrontation and polarization, and entrenchment of great power rivalry in Southeast Asia. While this may not be incorrect, it is important to recognize that both these developments would have occurred in any event. In the 1970s Vietnam had concluded a special relations agreement with Laos, and one with Cambodia in 1979, after installing its own client government. Had Thailand and ASEAN recognized the new regime in Pnom Penh, it would have been tantamount to accepting an Indochina "bloc" under the leadership of Hanoi. With international recognition, Hanoi would have consolidated its position with relative ease. Also plausible, but less so in my view, is the argument that without the opposition of ASEAN, Pnom Penh's dependence on Hanoi would have been much less, and in due course Cambodian nationalism would have reasserted itself, as has Laotian nationalism; this would have loosened Hanoi's control over Cambodia. This approach, especially from the point of view of Bangkok, would entail subjecting Thai national security to uncertain long-range factors over which it had no control. In retrospect, in the light of ongoing developments, it is possible to argue that ASEAN's Cambodia policy has not only contributed to Thai security but is, in fact, contributing to a loosening of Vietnamese control over Cambodia.

Similarly, the intensity of Sino-Vietnamese and Sino-Soviet conflicts was such that, with or without ASEAN, they would have imposed themselves on the Cambodian conflict. To the extent that ASEAN became a significant regional force, it was able to channel and modify external power involvement in support of its position and to contribute to stability in the region.

6. Conclusions and Future Prospects

ASEAN has certainly had a positive impact on peace and stability in Southeast Asia. Its contribution to civil order among the ASEAN states, to collective political defence in support of Thai security, and its stabilizing impact on the Southeast Asian strategic environment, are especially important. These functions are also the ones that have the potential for further development in the short term (three to five years).

Given that current cordial relations among many ASEAN states rest on a rather narrow base, the several outstanding disputes among them, and the residual mistrust and suspicion that continue to characterize a number of bilateral relationships, the civil order function in ASEAN will continue to be important. It should not be taken for granted. Every effort must be made to strengthen the political, economic, military-security, and socio-cultural foundations of the ASEAN community. ASEAN must become much more than a club of political leaders and a select few bureaucrats. In political terms, the relevance of ASEAN must extend to the national judiciaries, legislatures and the groups involved in interest aggregation and articulation. For this to happen, a higher level of cooperation is necessary. The core of ASEAN interest must grow and interlock the states and their societies at different levels and dimensions. ASEAN could also face a number of challenges in the coming years which could make the building of an ASEAN community even more difficult. These include leadership transitions, the disappearance or reduced relevance of the impetus for cooperation provided by the Cambodian conflict, emerging divergent role conceptions among the leading members of ASEAN, increasing attraction of Asia-Pacific economic cooperation for some members, possible emergence of "soft regionalism" in mainland Southeast Asia, and the possible expansion of ASEAN.

Simultaneous with strengthening the ASEAN community, every effort must be made to extend the civil order function to encompass the states of Indochina. Rapprochement among the major powers and the consequent reduction (but not elimination) of competition among them, and the distinct possibility that has emerged for a settlement in Cambodia, would appear propitious for this undertaking. This, however, would not be easy. The development of cooperative relations with the states of Indochina would be a major challenge for ASEAN, but one which it has explored since the early seventies. The agreement in 1973 among the foreign ministers had three components: convening of a Southeast Asian forum to discuss matters of mutual interest and concern; expanding ASEAN to include all states in Southeast Asia at an opportune moment; and improving bilateral relations with, and engaging in meaningful participation in, the rehabilitation and reconstruction of all the states of Indochina. This agreement also appears to have influenced the decision to leave open the 1976 Treaty of Amity and Cooperation for accession by all states in the region.

Expanding ASEAN deserves careful consideration and a cautious approach. ASEAN may not be ready as yet to accept new members, especially those with different socio-economic systems and which are also rather weak in political and economic foundations. Extension of the civil order function, however, does not necessarily have to be undertaken within the ASEAN framework. A new Southeast Asia forum can be established, as agreed to by the ASEAN foreign ministers in 1973, to discuss and agree upon measures to preserve and enhance stability in the region. The Jakarta Informal Meeting may in some ways be perceived as a prelude to such a forum. Vietnamese Foreign Minister Nguyen Co Thach has suggested some kind of a consultative body. Regular consultations and cooperation in such a forum can facilitate socialization, create trust and goodwill, and eventually pave the way for expansion of ASEAN.

ASEAN should encourage the development of bilateral economic relations between its members and the states of Indochina. Gradually, the latter countries could be drawn into ASEAN economic activities on a sectoral basis. In this connection, the recent Thai initiative to encourage economic cooperation among the mainland Southeast Asian economies should be viewed positively. ASEAN should also encourage the flow of international aid, trade and investment to the states of Indochina. Vietnam must in the process undertake to terminate its membership of COMECON.

The states of Indochina should be encouraged to accede to the 1976 Treaty. Simultaneously, attention must be devoted to maintaining stability in Cambodia. A settlement in Cambodia would not necessarily lead to stability in that country. Every effort must be made to ensure effective implementation of the letter and spirit of the settlement that is reached. Should the settlement break down and give way to a struggle for power among the four factions, unilateral external intervention must be prevented. Where necessary, collective action should be taken to contain the conflict to its internal dimension. Peace and stability in Cambodia is vital for peace and stability in Southeast Asia.

Another issue that merits urgent attention are the conflicting claims in the South China sea. In February 1989 there was an "accidental" clash between Chinese and Philippine naval forces, and in April the Chinese Navy occupied another islet previously held by Vietnam. The potential for further "accidental" clashes is ever present given that nearly every country is upgrading its maritime forces. Because of the many overlapping claims and the very different basis on which these claims are made, resolution of the territorial dimensions may be exceedingly difficult. It is, however, important to prevent military conflict and to guarantee safety of international navigation. There is a need for the institution of a multilateral framework to begin discussion of the many issues involved and, more importantly, to direct efforts away from the military measures to dialogue and negotiations on fundamental principles, to ensure stability in the area. It will be an irony if the ending of the Cambodian conflict is followed by tension and conflict in the South China Sea.

Fostering of cooperation among all states in the region (primarily through the extension of the civil order function) may also enhance the position of states in the region *vis-á-vis* external elements. ASEAN has demonstrated that collective action by regional states cannot be ignored. Cooperation among all states in the region can further enhance their influence in shaping their immediate strategic environment. They would, however, have to explore seriously the development of common policies and/or co-ordination of their positions in relation to China, whose strategic significance to the region would continue to increase, and in relation to Japan and the United States which would continue to be the major economic power for the region.

Southeast Asia may become more stable in conventional strategic terms, but the near total permeation and the consequent high dependence and vulnerability of the regional economies to international economic developments, over which they have little control, pose a new set of problems for countries in the region.

The Role of Regional Organizations in the Maintenance of Peace and Security in Latin America

Julio Carasales

1. The Building of the Latin American Regional Security System

Article 52 of the United Nations Charter not only does not forbid the existence and operation of "regional arrangements or agencies" to deal with "matters relating to the maintenance of peace and security" but urges Member States to "make every effort to achieve pacific settlement of local disputes through such regional arrangements" or "agencies" before referring them to the Security Council (paragraph 2). It goes on to state that the Council itself "shall encourage" such a way of settling peacefully local disputes (paragraph 3).

The positive contribution that regional organizations could make to the maintenance of international peace and security was already recognized in the Covenant of the League of Nations, whose Article 21 provided that "Nothing in this Covenant shall be deemed to affect the validity of international engagements, such as ... regional understandings like the Monroe Doctrine, for securing the maintenance of peace". (The drafters of the Covenant made the incredible mistake of identifying the Monroe Doctrine as a "regional understanding", when it was always a United States unilateral doctrine, as was pointed out by several Latin American States, among them Argentina on 28 February 1928).

The Charter of the United Nations includes another Article which has a bearing on the subject of this paper. Article 51, for which the Latin American delegations in San Francisco were among the strongest advocates, affirms "the inherent right of individual or collective self-defence if an armed attack occurs against a Member of the United Nations".

In the region known as Latin America, the regional arrangement or agency, as Article 52 of the Charter calls it, is the Organization of American States (OAS). Its name and Charter were adopted in 1948, three years after the establishment of the United Nations, but the regional organization was in fact much older, beginning with the First Inter-American Conference held in Washington in 1889–1890. A great number of meetings took place from that date onwards, and highly important documents and measures were adopted at them.

If one wished to go even further back in time, one could recall the famous Congress of Panama, held in 1826 on the inspiration of Simon Bolivar. (The Congress of Panama was not a Latin American conference, as is commonly believed. The United States was invited and appointed delegates, but, for various reasons, they were unable to attend). The Treaty of Union, League and Perpetual Confederation, signed on 15 July 1826, declared in its Article 2 that the purpose of the pact was

> "...*to support jointly, defensive and offensively if necessary, the sovereignty and independence of all and each of the confederate powers of America* ... *and to ensure, now and always, the benefits of an unalterable peace.*" Article 3 added that "the Contracting Parties bind and engage themselves to mutually defend each other against any attack which could jeopardize their political existence ..." [1]

In several Latin American Conferences, held during the Twentieth Century, different treaties were approved which initiated and developed principles embodied in the instruments currently in force in America relating to peace and security. I will not mention them, for the sake of brevity, but within the framework of the Inter-American Conferences, special reference should be made to the Conference for the Maintenance of Peace (Buenos Aires, 1936). On the proposal of the United States at the Conference, the American countries dealt in particular with these questions and adopted a Convention on the Maintenance, Strengthening and Reestablishment of Peace, thus laying the foundation for a system of consultation among the American States on how to face the outbreak of war *inside* or *outside* the continent. A Declaration of Principles on the Inter-American Solidarity and Cooperation, adopted at the same Conference, stated that "any act capable of endangering peace in America affects all and each" of the Nations of America.

The system of consultation established in Buenos Aires was refined at the Eighth Inter-American Conference (Lima, 1938), which decided that the consultations were to be carried out through Meetings of the American Foreign Ministers (Lima Declaration, resolution CIX, operative paragraph 4). The Second World War, which began the following year, was the cause of the first three Meetings of Consultation of Foreign Ministers (Panama, 1939; Havana, 1940; and Rio de Janeiro, 1942).

In 1945, the American Conference on Problems of War and Peace approved the so-called "Act of Chapultepec" (8 March 1945). Part I, paragraph 3, declared that

> "...*every attack of a State against the integrity or the inviolability of the territory, or against the sovereignty or political independence of an American State shall* ... *be considered as an act of aggression against the other States which sign this Act.*"

And Part II recommended the conclusion of a treaty for the purpose of "meeting threats or acts of aggression against any American Republic". The envisaged treaty was concluded in 1947 in Rio de Janeiro, Brazil, and is known as the Rio Treaty, but its full name is the Inter-American Treaty of Reciprocal Assistance (TIAR).

The preceding sketchy summary of the development of regional activities in America in the field of peace and security attempt to show that those activities had a long history and preceded the establishment of the League of Nations and the United Nations, even if the two regional instruments now in force in the region, the Rio Treaty and the OAS Charter, were adopted after 1945.

2. The Main Features of the Regional Peace and Security System

The Organization of American States has been called "the oldest and most prestigious regional security system in the world" [2]. Perhaps not everybody will agree with this assessment, but it is a fact that Latin America, compared with other regions of the world, has a highly positive record in the field of peace and security. To what extent that record is due to the operation of a reasonably successful regional security system is a matter of opinion, but certainly America does not lack a fair amount of disputes, some very serious ones, affecting its members, and it cannot be denied that very rarely had those disputes been the origin of hostilities or, even more rarely, of open war. I think that this moderately satisfying state of affairs can be attributed, as with almost everything in the international field, to the political will of the states concerned, which have generally shown a remarkable restraint in the peaceful management of their disputes. But it would be unfair to ignore the positive contribution which the existence and availability of an adequate regional machinery and procedures can and did make to the maintenance of peace and security in America.

Some comments seem to be in order. First, I have been using in this paper the terms "America" and "Latin America" almost as synonymous, which certainly they are not. But we have to face the fact that, for the maintenance of peace and security in Latin America, the competent regional organization is not a Latin American but an Inter-American institution, the Organization of American States, of which a non-Latin American State, and certainly the most powerful one, the United States, is a founding member.

The presence of a non-Latin member amongst a great majority of Latin American States has been, from the very beginnings in 1889, a very important problem in the regional organization. The advantages and disadvantages of such a situation, which derive not so much from the differences of racial origin or language as from the enormous disproportion of power between the one and all the others put together, has been the subject of a long and unfinished debate in the region. But the fact of the matter is that OAS is *the* appropriate international regional organization to deal with peace in Latin America and that a most important member of that organization is a non-regional state (if we consider Latin America as the region).

Another factor that we have to take into account is that OAS grew and became what it is because it congregated a group of countries with the same or very similar racial origin, language, cultural background and political history. The only non-Latin member, the United States, shared many of the same political values and beliefs. That made OAS a predominantly, if not a purely, Latin American institution. This common heritage of the OAS founding members ceased to be a reality when, from the early 1960s, new States began to appear in America, almost all of them former British colonies in the Caribbean area, English-speaking and with a non-Latin population. The great majority of them became members of OAS. Today they constitute approximately a third of OAS membership, making it a somewhat different organization. How much different, is a debatable subject. But what is more pertinent is that none of them (except Trinidad-Tobago) ratified the Rio Treaty, and, consequently, they remain outside the operation of one of the pillars of the regional security system. The OAS discussed at length whether to make or not an indispensable condition to be admitted to the OAS the ratification not only of the OAS Charter but also of the Rio Treaty. The decision was in the end negative and this was considered by many a mistake, an opinion shared by me.

Another comment is appropriate. In practical, if not legal, terms, in the vast area known as Latin America, a distinction could and should be made in the field of the maintenance of peace and security, between two sub-regions: South America, and Central American and the Caribbean. South America, by far the largest and more populated area, has a generally peaceful history in this century. No major war has taken place between South American countries since the signing of the Rio Treaty and the OAS Charter. The Malvinas War of 1982 was an exceptional event and involved a non-American power as a belligerent. Very few of the meetings held under the OAS Charter or the Rio Treaty, to deal with threats to peace in the region, had a South American origin or theatre.

Completely different is the case of the other sub-region: Central America and the Caribbean. This area has been and continues to be the arena of political upheaval, border conflicts, civil wars and open hostilities, which, in this zone, tend to have a major international ingredient and even an element of superpower confrontation not present in the other part of the continent. Be that as it may, it is a fact that almost all the Meetings of Consultation held according to the OAS Charter or to the Rio Treaty, and almost all, if not all, the meetings held by the Permanent Council of the OAS (or its preceding body, the Council of the OAS, until the 1967 Charter revision changed its name and some of its functions) acting as provisional Organ of Consultation (according to Article 12 of the Rio Treaty and Article 82 of the Charter) had to deal, in this field, with problems related directly or indirectly to Central America and the Caribbean. It seems that the smaller and weaker the countries of a particular area, the greater is the temptation of foreign powers to intervene and, consequently, the complexities of each situation or dispute.

I will now describe succinctly the system established within the OAS framework to deal with the maintenance of peace and security in the region and the peaceful settlement of international disputes. The system was intended to rest on three bases: the Rio Treaty, the OAS Charter and the American Treaty of Pacific Settlement or Pact of Bogota, signed together with the Charter on 30 April 1948. But this last Treaty was ratified by few American States and remained, in practical terms, outside the operation of the system.

The Rio Treaty was drafted and signed in 1947, in compliance with the Act of Chapultepec and a year before the adoption of the OAS Charter. Two assumptions were made by the authors: first, that all members of the regional organization would become parties to the Treaty, and, secondly, that for the operation of the Treaty use could be made, consequently, of organs of the regional organization. That is why the body through which the Rio Treaty becomes operative is not a new or an *ad hoc* body, but the Meeting of Consultation of Ministers of Foreign Affairs, established in Lima in 1938. Until such a Meeting takes place, the Governing Board of the Panamerican Union (later the Council) may act provisionally as Organ of Consultation (Article 12).

The Rio Treaty condemns war and the use or threat of force (Article 1); includes the undertaking by the parties to settle every controversy between them by peaceful means (Article 2); declares that an armed attack by any State against an American State shall be considered as an attack against all (Article 3); defines a geographical area to which the Treaty is applicable (Article 4); provides for the operation of the Treaty in a broad set of circumstances, which include non-armed attacks or "any other fact or situation that might endanger the peace of America" (Article 6) and enables the Organ to "take in addition all other necessary measures to re-establish or maintain inter-American peace and security and for the solution of the conflict by peaceful means" (Article 7). Article 8 enumerates the kind of measures that the Organ of Consultation could take.

It is important to stress that the Rio Treaty has a *double function*: *to defend* an American state which has been attacked by an extra-Treaty aggressor and to do the same when the aggressor is an American State. At the same time, another purpose of the Treaty is *to maintain peace* in America and to help to solve peacefully disputes between American States. In fact, the Rio Treaty has been used more often for the second function than for the first, and, as noted before, for events connected mainly with the Caribbean and Central America.

The second step in the establishment of the regional security machinery was taken in 1948 in Bogota, Colombia, during the IX Inter-American Conference. There, the regional organization fulfilled a long-sought goal: to have a Charter. (The Bogota Charter was fully revised in 1967 in Buenos Aires and in 1985 in Cartagena, Colombia. Both sets of amendments entered into force some time later in each case, the last one on 16 November 1988).

Chapter VI of the OAS Charter, entitled "Collective security", reproduced in its Articles 27 and 28 some of the fundamental provisions of the Rio Treaty and made a reference to "the special treaties on the subject", meaning obviously the

Rio Treaty. When Part Two of the Charter referred to the Organs of the OAS, it naturally included among them the Meeting of Consultation of Ministers of Foreign Affairs, to consider problems of an urgent nature and of common interest to the American States, and *to serve as the Organ of Consultation*, that is, as a body of the Rio Treaty (my emphasis).

It is to be noted that the Meeting of Consultation of Foreign Ministers could take place outside the Rio Treaty, to consider "problems of an urgent nature and of common interest". Even if a great number of issues could be included in the preceding wording, the fact is that almost all meetings of this kind were held to deal with questions pertaining in one way or another to peace and security. The reasons for choosing in those cases a Meeting of Consultation according to the Charter, instead of according to the Rio Treaty, would have to be discussed with reference to each separate occasion. In general, it would seem that the phrase of the Charter (Article 39 in 1948, which became Article 59 in the Buenos Aires amendments of 1967 and Article 60 in the Cartegena amendments of 1985), avoids the characterization of the problem considered as a threat to international peace and security, as recourse to the Rio Treaty would imply, and, consequently, the possibility of the adoption of mandatory sanctions as provided for in that Treaty.

The Charter also gives important functions in the area of the maintenance of peace to its only permanent organ besides the General Secretariat, the Permanent Council. In 1967 this replaced its predecessor, the Council, which in turn replaced in the Charter of 1948 the Governing Board of the Panamerican Union. It should be noted that the change was not only of name but also of functions and competence. The Permanent Council could act under two capacities. First, as a provisional Organ of Consultation, according to Article 12 of the Rio Treaty and Article 82 of the present Charter. The original intention was to enable the Contracting Parties to act immediately in an emergency, while the Foreign Ministers were gathering. But later it was shown that the actions taken by the Council were sufficient, and made the Meeting of Foreign Ministers unnecessary. The consequence was that the Council often met under the Rio Treaty and either made a symbolic call for a Meeting of Foreign Ministers to be held at an undetermined date (with the understanding that it would never take place), or acted simply as a provisional Organ of Consultation, because everybody thought it preferable not to magnify the issue, and keep it at the level of Permanent Ambassadors to the OAS. (All OAS Member States are members of the Permanent Council).

It should be noted that when the Permanent Council is acting as provisional Organ of Consultation the delegates of countries which are not parties to the Rio Treaty cannot vote. The situation thus created was not easy to manage and gave rise to different interpretations, which the 1985 amendment of Article 81 (now 82) clarified completely. Almost a third of the members of the Council could not take an active part in the performance of that organ, and that was certainly anomalous. It proved that the decision taken in the 1960s to accept the existence of members of the OAS not parties to the Rio Treaty was inconsistent with the text of the Charter and of the Treaty, both of which should have – to avoid such an inconsistency – been amended to bring them into line with the new state of affairs.

Besides acting as a provisional Organ of Consultation on the basis of the Rio Treaty, the Permanent Council of the OAS has a responsibility of its own, entrusted to it by Article 83 (formerly 82) of the Charter: "The Permanent Council shall keep vigilance over the maintenance of friendly relations among the Member States, and for that purpose shall effectively assist them in the peaceful settlement of their disputes."

The next articles of the Charter specified the requirements, procedures and powers of the Council to act in this field. The Council should always proceed at the request of a concerned party but Article 116 of the 1985 Charter gives the Secretary-General of the OAS the right – which he did not have before – to "bring to the attention of the General Assembly or the Permanent Council any matter which in his opinion might threaten the peace and security of the hemisphere", reproducing the language of Article 99 of the UN Charter.

Until the Charter revision of 1967, the Council was assisted in the discharge of this mission by an Inter-American Peace Committee, which in the Protocol of Amendments of 1967 was replaced by an Inter-American Committee on Peaceful Settlement. This second body was less and less utilized, mainly because the political situation changed in America, but also because its membership was rarely the most adequate to deal with a particular case. The function of ascertaining the facts and of offering good offices in a dispute requires a careful and balanced selection of the countries who will perform those tasks, a selection to be made in close consultation with the parties to the dispute. This condition was not normally met by a Committee whose composition had been decided beforehand by elections held at regular intervals. These realities help to explain why, in the Cartegena amendments of 1985, the Inter-American Committee on Peaceful Settlement was abolished and the Permanent Council was given the right, by a new Article 85, to establish, with the consent of the parties to the dispute, *ad hoc* committees appropriate to each case.

Article 2 of the OAS Charter proclaims as "essential purposes" of the Organization: "(a) to strengthen the peace and security of the continent"; "(c) to prevent possible causes of difficulties and to ensure the pacific settlement of disputes that may arise among the Members States"; "(d) to provide for common action on the part of those States in the event of aggression"; and "(e) to seek the solution of political, juridical and economic problems that may arise among them". Among the Principles on which the OAS is based, set forth in Article 3, are the following:"(a) international relations"; "(f) the American States condemn war of aggression ... " and "(g) an act of aggression against one American State is an act of aggression against all the other American States".

Chapter V of the Charter refers to the "Pacific Settlement of Disputes" and Article 23 states that "international disputes between Member States shall be submitted to the peaceful procedures set forth in this Charter", which include the traditional ones such as direct negotiation and good offices (Article 24). Before the 1985 Charter revision, Article 23, continued with the phrase: "before being referred to the Security Council of the Nations". This clause gave rise to a lengthy debate, repeated several times, about the nature of the obligation of all American States to submit their differences to the regional organization, as a prerequisite to doing so at the UN Security Council. It should be recalled that Article 52, paragraph 3 of the UN Charter compels the Security Council "to encourage the development of pacific settlement of local disputes through such regional arrangements or by such agencies".

When, on this basis, the consideration by the Security Council of cases submitted directly to it was opposed by some countries, the answer was that no United Nations member should have less rights than any other just because it belonged to a regional organization, and that a state should always have the right to raise its case in the Security Council if, for any reason, it feared that it would not receive a fair hearing in the regional agency. The debate took place both in the United Nations and in the OAS. The issue was definitely settled in 1985 when the last phrase of Article 23 was eliminated and a new paragraph added giving special recognition to "the rights and obligations of the Member States under Articles 34 and 35 of the Charter of the United Nations". Hence there is now absolutely no possible collision between the rules of the world and of the regional organizations concerning the maintenance of peace and security in Latin America. Both have a role to play and both have played it.

3. The Record of the OAS and its Present Prospects

If we look at the record since the United Nations was established, we can see that the OAS has been fairly successful. It is beyond the scope of this paper to deal with all the threats to, or breaches of, peace or international disputes which have taken place in Latin America during these last four decades, and with the manner in which the regional security system reacted in each case. In a few cases the issues were not even submitted to the OAS but, in general, the regional bodies dealt with them in one way or another. In some cases, the regional system solved them, in others, it made a more or less positive contribution to the settlement, and in still others it failed. But it can be fairly said that its record is not worse but very probably better, perhaps much better, than the record of other comparable international organizations, beginning with the United Nations.

The preceding sentence is not intended to imply that the OAS system does not have its shortcomings. What is more serious, there have been in recent years several episodes which have had negative repercussions for the regional organization [3]. The following are a few examples:

The South Atlantic War. The Rio Treaty was invoked by Argentina and the measures finally taken by the Twentieth Meeting of Consultation were considered by many to be very modest if not bland. The explanations could be several and the question has been widely discussed4. But the result has been that, rightly or wrongly, the value of the Rio Treaty has greatly diminished, its usefulness as a security arrangement has seriously deteriorated, and its future as a source of regional security and peaceful settlement of disputes does not seem, at least at present, promising.

The Grenada Invasion. This action was taken completely outside the OAS system, making use of a little known sub-regional Treaty (the Treaty of the Organization of Eastern Caribbean States, whose members were Antigua-Barbuda, Dominica, Grenada, St Kitts-Nevis, St Lucia, St Vincent-Grenadines and Montserrat) and in blatant disregard of OAS principles, particularly non-intervention. One should also not ignore the many instances of less open intervention by an American state in the affairs of another, especially through infiltration or support of local guerrilla movements. Many think that the OAS has not reacted to this kind of intervention in the manner it deserved.

The blow to the OAS regime that the Grenada invasion signified was compounded by the very weak reaction to it adopted by the competent OAS bodies. Even if among English-speaking members of the OAS the positions regarding the Grenada case were widely divergent, this episode contributed to reinforcing one of the consequences of the Malvinas war: the feeling that the OAS was no longer a gathering of homogenous countries but is now the aggregate of two different groups, one speaking English and the other speaking Latin languages, with dissimilar views and attitudes that go beyond the idiomatic disparity.

The Central American Crisis. Central America and the Caribbean were always the theatre of the most active OAS participation, but this is not so in the case of the present situation. International action for the solution of the crisis has not been in the hands of an OAS organ, but of an outside body, the Contadora Group, composed of the four foreign governments outside the sub-region most closely interested in the conflict: Colombia, Mexico, Panama and Venezuela. Other countries of Latin America (Argentina, Brazil, Peru and Uruguay) are helping through the Support Group of Contadora.

Why in this case the OAS machinery has not been used would require a special paper, but the fact is that the regional security system has been largely ignored in dealing with this crisis. The OAS had no choice but to endorse the activities undertaken by Contadora and the bodies and meetings operating under its umbrella. Whatever the reasons for setting aside the OAS, the results certainly do not favour the credibility of, or increase confidence in, the traditional regional institutions.

All this being said, the perennial truth remains: in principle, international organizations are neither good nor bad in themselves; everything rests on the political will of their Member States. The inter-American security system is not passing through one of its best periods. Only the future will tell if this situation will continue or even become worse, or if, on the contrary, the countries of the region will again make ample use of the institutions which they have developed, after a long and sometimes difficult process, to deal with the maintenance of peace and security in the area.

The machinery is there. The principles, rules and procedures are there. These could perhaps be improved upon, because nothing is perfect, but basically they are sound and no major change seems necessary, at least in my opinion. If the countries of the region decide to utilize them as in the past, the machinery is available. If they decide to seek new ways to ensure peace and security, at least in some particular cases, there is no problem. But whatever the reason for doing so, it would not be for lack of regional means already at their disposal.

What is really important, in any case, is the will of the American States to maintain peace and security in the region and to settle peacefully their disputes, making full use or not of their regional agency, as they wish. And that will, in my view, and with all the qualifications and allowances that could be made, fortunately has existed and continues to exist in Latin America.

References

1 See Carnegie Endowment for International Peace *Conferencias Internacionales Americanas 1889–1936*. (Washington, 1938) p. xxix. It should be noted that this publication, a compilation of documents adopted since the First Inter-American Conference, is meaningfully preceded by documents relating to the Congress of Panama

2 J. N. Moore, "The OAS and the struggle for peace: The Central American conflict as a case study", in *Anuario Juridico Interamericano, 1986* (Washington DC, 1986)

3 See J. Child, "Present trends in the Inter-American Security System and the role of the Rio Treaty" in *Anuario Juridico Interamericano, 1983* (Washington DC, 1983) pp. 43–82. The picture shown in this article had not changed substantially six years later

4 A. Quiroga, "El caso de las Malvinas visto desde la perspectiva del Organo de Consulta del TIAR", in *Anuario Juridico Interamericano, 1983*, pp 223–93

Part VII

Energy – Environment – Development Interaction

Lifestyles, Energy Choices and Development

José Goldemberg

1. Introduction

One of the serious problems threatening internal and external security of developing countries is the competition for raw materials and energy. Although energy is not the only, nor even the most important, problem in developing countries, strong connections can be made between energy consumption, economic well-being, and security.

One of the most striking features of the present global distribution of energy use is the enormous disparity between the developing and industrialized countries. The 70 per cent of humanity that lives in developing countries accounts for only 30 per cent of global energy consumption. In fact, if only oil, natural gas, coal, and electricity – the so-called "commercial" energy sources – are considered, the developing countries' share of global energy shrinks to less than 20 per cent. The disparity is even more striking on a "per capita" basis. On average, a person from a developing country consumes less than one-sixth of the energy consumed by a person from an industrialized country, and less than one-tenth as much as the average US citizen. Since the efficiencies of energy use are also extremely poor in developing countries, the low levels of energy use correspond to abysmal levels of services provided by energy.

Despite some diversity of interest among developing countries, they have strong common interests. They are poorer than developed countries, many have been colonies, and the international economic context in which they must operate is dominated by the industrialized countries. And importantly, in terms of this analysis, most developing countries are highly stratified into "dual societies", with rich elites and poor masses. This polarization creates similar patterns of energy consumption among developing countries.

2. "Dual Societies" and Elites

A wide chasm of incomes, consumption patterns, attitudes, aspirations and lifestyles separates the elites of developing countries from the masses. The elite – industrialists, commercial traders, landlords, government officials and bureaucrats, upper echelons of the armed forces, professionals and some skilled craftsmen – constitute only about 10 per cent of the population of these societies, but to them goes the lion's share of these societies' economic rewards. And despite their small numbers, the elites are politically powerful, controlling virtually all aspects of the decision-making process in these countries. Their conspicuous affluence contrasts markedly with the abject poverty of the politically-weak masses, who are dispersed in the villages of the countryside and concentrated in the slums of the metropolises. In short, developing countries consist of tiny islands of affluent splendour amidst vast oceans of miserable poverty. The consumption patterns and lifestyles of the elites of developing countries are strongly influenced by those in the industrialized countries. It is as if the elites in developing countries practice a philosophy best described thus: all that is rural is bad, all that is urban is better, and all that is current in the industrialized countries is best.

The stratification of these societies into elites and poor masses is the fundamental reality which has shaped the current energy systems of these countries. To ignore these realities, as all global energy strategies have done, and as international organizations and agencies unfortunately are obliged to do (on the grounds of non-interference in internal affairs), is to miss the essence of the situation. It is crucial, therefore, to explore further the impact of "dual societies" and their elites on the energy systems of developing countries.

3. The Elites and the "Modern" Sector of Economies in Developing Countries

The elites of developing countries in their desire for status symbols and modern conveniences seek goods and services similar, if not identical, to those available in the industrialized countries. Initially, this desire can be satisfied by imports from the industrialized world. But such imports cannot be sustained for long, because they result in balance-of-payments problems; a local productive base is required. A so-called "modern" sector, involving import-substitution and modernization of the industrial, agricultural, transport, commercial and residential sectors, is necessary. Of course, there are marked differences in the extent to which modernization of these sectors has taken place in different developing countries. In some of the countries, for example, Brazil and India, the modernization started several decades ago; in others, for example, Tanzania and Indonesia, it is a recent phenomenon. Generally, though, this modernization had tended to be along the lines found in industrialized countries.

In industry, the goal is to establish and/or increase the production of those basic materials and products for which the developing country has resources, for example, iron and steel, cement, textiles, sugar and aluminium, and other comparative advantages. These objectives are usually achieved with imports of technology from the industrialized countries – hence, the industrial sectors in developing countries are embryonic versions of the corresponding sectors in the industrialized countries, even retracing their course of development and repeating their mistakes – for example, their dependence on petroleum.

In agriculture, the aim is to overcome the limitations of traditional agriculture and its low productivity (arising from its dependence on human labour, draft-animal power and rain-fed irrigation), by adopting the modern practices of mechanized ploughing, hybrid seeds, irrigation upon demand, fertilizer applications, crop care, harvesting and post-harvest cropprocessing.

In the transport sector, the main thrust is towards transportation modes fuelled by petroleum products. This implies use of trucks for freight, and buses, automobiles and aeroplanes for passenger transport. In the commercial sector, the growth of modern business firms, particularly the local branches of transnational corporations, is leading to office buildings and practices similar to those found in the industrialized countries – hence the presence of illuminations and space-cooling in cities in developing countries. In the residential sector, the trend is towards modern energy carriers, such as electricity and liquified petroleum gas for cooking, electricity for lighting, and the use of electrical appliances such as refrigerators, electric fans and air conditioners.

This process of modernization has led to a pattern of energy use in the modern sector of developing countries which is quite similar to the energy use pattern in the industrialized countries. The distribution of consumption of *commercial* energy in developing countries is not too different from that in the industrialized countries, the main difference being the much lower share of energy use for commercial buildings in developing countries. And it has been more convenient to use the same modern commercial energy carriers – electricity, oil, gas and coal – as are used in the industrialized world, rather than traditional biomass fuels.

In sum, the demand of elites in developing countries for a modern sector, to satisfy their desire for the goods and services of the industrialized countries, has generated a similar pattern of consumption of commercial fuels. In contrast, the poor in developing countries must rely on traditional biomass energy sources, and their pattern of energy use is totally different.

4. The Poor and Non-Commercial Energy

The poor in developing countries may be broken down into two groups:

- the urban poor who inhabit the slums, shanty-towns, barios and favelas of developing countries' metropolises;
- the rural poor who live in the countryside.

In a country like India, with 76 per cent of the population in rural areas, the rural and urban poor can be considered to constitute about 49 per cent and 5 per cent respectively of the total population, that is the rural poor are roughly ten times more numerous than the urban poor.

The urban poor in the developing world benefit far more from commercial energy and modernization than the rural poor. These benefits involve the use of modern fuels for cooking, electricity for lighting, and piped hot water either to the homes or to nearby locations. In contrast, the rural poor in most developing countries benefit only marginally from modernization and the energy carriers associated with it. The rural electrification that is widely emphasized in developing countries does not necessarily lead to the electrification of all homes. With only a small percentage, say 10–15 per cent, of homes in electrified villages being able to afford domestic electrical connections, the poor have no choice other than depending on oil lamps for lighting. Vegetable oils used to be the main source of illumination, but they have been largely replaced by kerosene over the past few decades. Even this small improvement has not affected cooking. The vast majority of the rural population in developing countries still depends mainly on biomass sources – chiefly firewood, but also agricultural wastes and cattle-dung cakes. These sources are known as "non-commercial" sources because, instead of being purchased, they are usually gathered in the form of branches, twigs and the like, often with considerable effort by women and children.

Thus, commercial energy, in the form of electricity, liquified petroleum gas and petroleum, plays only a minor role in the life of the rural poor. Since a large percentage of the population of most developing countries is rural, this means that non-commercial energy plays a very significant role in the energy systems of these countries. On average, non-commercial energy constitutes about half the total energy used in a developing country. But it was only during the last decade that the importance of non-commercial energy use has been recognized and studied.

5. The Unviability of Present Energy Consumption Patterns in Developing Countries

The patterns of energy consumption referred to above have led to three different energy crises in developing countries:

- an oil crisis (in oil-importing developing countries);
- a fuelwood crisis;
- an electricity crisis (in some countries).

The oil crisis, which is the best known, is manifested in a rapidly worsening balance of payments situation with large fractions of the export earnings of developing countries going on oil imports. As a result, the economies of most oil-importing developing countries are in trouble. Moreover, the unpredictability of future oil prices, and therefore of the impacts of these prices on internal economies, has made economic planning extremely difficult.

The crisis of non-commercial energy, often referred to as "the other energy crisis", not only involves severe scarcities of cooking fuel, but also the associated problems of tree-felling, deforestation, soil erosion and desertification.

The electricity crisis is primarily associated with the growth of the industrial sector in developing countries, and its rising demand for electricity. On the other hand, there are countries where hydroelectric and coal resources are scarce, and where oil is used for electricity generation. In these countries, the increases in oil prices are directly affecting electricity generation, throttling the growth of industrial production and forcing an increase in imports of products which could otherwise have been locally manufactured. In these countries, the electricity crisis is thus only a manifestation of the oil crisis. On the other hand, even in hydro-rich countries, the rapid influx of power-intensive industries, particularly electrometallurgical industries, such as aluminium, can throw the electricity supply-demand situation out of balance. The point is that power projects have long gestation times compared to the gestation times of industries and, at low levels of industrialization and total electricity consumption, a few power-intensive projects can drastically outstrip supply. Brief periods of electricity excess are likely to be followed by periods of shortage, creating a problem which is more one of energy planning than of energy resources. But identifying the problem as a planning problem does not in any way diminish its intensity and impact.

The oil-fuelwood and electricity crises are, in fact, inter-related. They originate, ultimately, from the basic tendency in "dual societies" to concentrate economic growth on the needs of the elite, and to ignore or under-emphasize the needs of the poor. This bias has led to the creation in developing countries of energy-norms which may have been economical in the industrialized countries during the era of cheap energy but are uneconomic today in both the developing and the industrialized worlds. Most developing countries have concentrated their energy planning efforts on electricity and oil, and consequently, the firewood problem has been neglected, even though it plays a crucial role in the lives of the poor.

The time has come, however, when the patterns of energy consumption in the developing countries cannot be sustained for long because the elites are as much threatened by the oil crisis, and, in many countries, by the electricity crisis, as the poor are by the fuelwood crisis. The adoption of alternative energy strategies has thus become a matter of immediate necessity and long-term survival for both the poor and the rich.

6. Energy for Development

It has been stressed over and over again that energy in developing countries must become an instrument for development. This means that energy must become an instrument for satisfying basic human needs. Hence it is important to estimate how much energy is required per capita for this purpose in order to get an idea whether the provision of this energy is feasible. A rigorous estimation of the per capita energy requirements for the satisfaction of basic human needs is extremely difficult. Nonetheless, an estimate is essential because it permits a preliminary assessment of the global requirements of energy, if developing countries set out to satisfy basic needs as a principal development objective.

A systematic approach to the quantification of direct energy needs may involve the following steps:

- listing the various energy needs such as cooking, food preservation, lighting, space heating, space cooling, water pumping, water heating, recreation and social communication;
- defining minimum physical targets, such as temperature and illumination levels;
- estimating the useful energy required for the attainment of these targets taking into account climatic conditions;
- calculating the direct primary energy needs from useful energy requirements, assuming efficiency values for the technologies of energy conversion, distribution and enduse, as well as conventional energy carriers.

The results of such computation, using average present-day technologies for hot (25°C), temperate (18°C), and cold (10°C) climates, lead to direct energy needs of 0.3–0.8 kW per capita (see Table 1).

Table 1. Calculation of Direct Energy Needs for Various Climatic Regions

Basic Need	Per Capita Requirement (kW)		
	Hot Climate	Temperate Climate	Cold Climate
Lighting	0.06	0.07	0.11
Space Heating	–	0.09	0.34
Space Cooling	0.02	0.01	–
Food Preservation	0.05	0.04	–
Cooking	0.07	0.09	0.11
Hot Water	0.04	0.09	0.13
Leisure, etc.	0.05	0.05	0.07
Total	0.29	0.44	0.76

Source: V. Bravo, G. Gallo Medoza, J. Legisa, C. E. Juarez and I. Zynierman, *Estudio sobre Requerimientos Futuros de Fuentes Non Convencionales de Energia en America Latina* (Fundacion Bariloche, Buenos Aires, 1979).

The indirect energy embodied in all of the goods and services needed for the satisfaction of basic needs (though not in conspicuous consumption) have, however, to be taken into account. This requires that the per capita energy consumption and the quality of life be correlated. Then the question arises how the quality of life can be quantified. In this context, an indicator known as the Physical Quality of Life Index (PQLI) has been used. The PQLI is a composite index which gives equal weights to indices for life expectancy, the infant mortality rate, and literacy. These indices are assigned values on the scale of 1 to 100, with the lowest and highest values corresponding to the worst and best world performances, respectively. When the PQLI indices of various countries are plotted against their per capita energy consumption (see Figure 1), it is observed that above the range of 1.1 to 1.3 kW per capita, the improvements in PQLI with increasing energy consumption are only marginal. Hence, it can be concluded that, if there is continued use of the energy technologies currently in vogue in these countries, an energy consumption rate of 1.1 to 1.3 kW per capita is required to satisfy basic human needs.

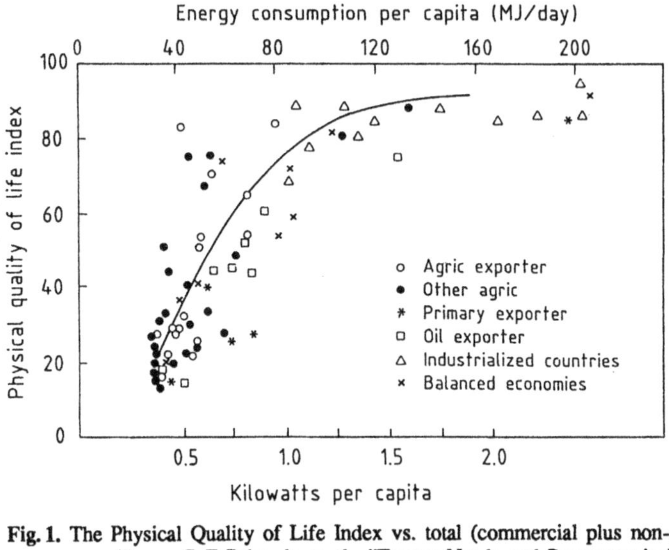

Fig. 1. The Physical Quality of Life Index vs. total (commercial plus non-commercial) per capita energy use. (Source:P. F. Palmedo et al., "Energy Needs and Resources in Developing Countries", Brookhaven National Laboratory Report No. BLN 50784, March 1978.)

7. Conclusion

It seems to be possible to supply the basic energy needs of the people living in developing countries with an amount of energy equivalent to 1 kW of power from commercial and non-commercial sources. Moreover if this amount of power is used efficiently it can be demonstrated [1] that it will suffice to sustain a level of amenities similar to that of European countries in the mid-1970s. This, of course, would require the introduction of modern – but current – technologies in the use of energy.

The use of modern efficient methods of converting primary energy (coal, oil, gas, hydro and solar) could, therefore, solve the energy problems of the developing countries, without a repetition of the costly mistakes made in the past by the industrialized countries, and without the use of immense resources to sustain wasteful energy consumption patterns.

References

1 J. Goldemberg, T. N. Johansson, A. K. N. Reddy and R. H. Williams, *Energy for a Sustainable World* (New Delhi, India, 1988); and J. Goldemberg, T. N. Johansson, A. K. N. Reddy, and Robert H. Williams, "Basic Needs and Much More With One Kilowatt Per Capita", in *Ambio*, 4, No 4–5 (1985)

Broadening the Global Security Debate: Regional and Global Dialogues on Environmental Issues

Irving Mintzer

1. Environmental Degradation Threatens Global Security

Evidence is mounting in many countries that traditional, economically-important activities are posing increasing threats to the environment. The stresses are being felt simultaneously at the regional, national and global level. These impacts of human activity take many forms. Damage from the loss of forests and the acidification of lakes and streams are proving a major cost to economies in Europe, the United States and Canada. Dangerous levels of water pollution from agricultural and industrial chemicals can be measured from China to California, from Poland to the Philippines. Soil erosion threatens agricultural productivity from Iowa to India. Tropical deforestation in Asia, Latin America and Africa increases the risk of soil erosion at the same time that it reduces forest productivity and biological diversity.

Fossil fuel combustion and biomass burning are among the principal culprits, releasing billions of tonnes of carbon and nitrogen compounds into the atmosphere each year. Some of these compounds, including carbon dioxide, methane and nitrous oxide, have been building up in the atmosphere for a century or more. Their continued build up threatens to alter rapidly global and regional climate regimes. Other combustion-related releases, including sulphur oxides and odd oxides of nitrogen, although short-lived in the atmosphere, tend to increase the risks of acid deposition and oxidant build up. Releases of chlorine-containing industrial compounds, including carbon-tetrachloride, methyl chloroform, methylene chloride and the chlorofluoro-carbons (CFCs), add to the risks of both ozone depletion and global warming.

As a result of these effluent releases, the composition and behaviour of the atmosphere is changing. If current trends continue, the combined impact of these changes will be to place severe stresses simultaneously on many of the natural processes that support economic growth and development. For example, global climate change, oxidant build up and soil erosion could combine to reduce the productivity of agricultural and forest systems during a period of rapid global population growth, when more, not less, food, fodder and fuel will be needed. Ozone depletion, wetland loss and the runoff of toxic agricultural chemicals may reduce the protein production of marine fisheries in the next several decades, just at the time when additional sources of cheap protein will be required to feed the growing populations of the Third World. Uncontrolled biomass burning and oxidant build up threaten the health and strength of billions, especially women and children in the Third World.

The net effect of these linked environmental problems is to place severe stresses on human societies at a time of increasingly urgent needs and rising intra- and inter-regional tensions. If ignored today, these regional and global environmental stresses could produce a generation of environmental refugees – hapless, homeless, often destitute, people who will be forced from one country to another with no visible means of support.

Any one of these stresses might be enough to increase political tensions among nations. The climate change resulting from the build up of greenhouse gases, for example, is expected to produce a sea-level rise of 0.4–1.2 metres during the next century. A rise of one metre would flood about 15 per cent of the agricultural land in Egypt, an area which currently supports 20 per cent of the population and produces more than 20 per cent of the country's Gross National Product. In the Bight of Bengal, a one-metre sea-level rise could leave 30 million homeless in Bangladesh alone, even before the effects of more severe tropical storms are taken into account. During a period of rapid population growth in both India and Bangladesh, it is not clear that 30 million refugees would be warmly welcomed upon their arrival in India.

But the risks of environmental degradation are not limited to Third World countries. A change in regional climate due to the greenhouse effect could reduce the availability of water in the Colorado River in the United States. Any such reduction would increase political tensions within the United States where the average annual flow of the Colorado is already overcommitted by legal obligations. If flow is reduced by climate change, the continued economic growth of Los Angeles might have to be evaluated against the water demands of less populous regions of the country. If the United States tried to increase withdrawals of Colorado River water at the expense of Mexico, its neighbour to the South, the resulting economic losses in that country might increase the already high level of bilateral tension over illegal immigration.

Examples of such stresses and the risks they entail abound across the globe. If the rate of environmental degradation continues to increase, it is possible that the resulting pressures will undermine the stability of civilization as we now know it. The resulting pressures on geopolitical relationships may be impossible to reduce through the mechanisms of conventional arms control negotiations and confidence-building measures.

2. The Political Atmosphere is Changing

Fortunately, at least in some quarters, the political atmosphere relating to these problems is changing. A growing number of international statesmen are pointing out the problems of global environmental degradation and recognizing the threat that this poses for international peace and economic growth. In 1987, the UN World Commission on Environment and Development (the Bruntland Commission) focused public attention on the linkage between environmental protection and sustainable development. The Commission report, *Our Common Future*, noted that we no longer face a choice between environmental quality and economic growth. It concluded that sustainable economic growth was only possible if the economically-important services provided by natural ecosystems were maintained.

Political leaders in both North and South have recently accepted the visions of Mahatma Gandhi: that the environment is the foundation of economic growth. In June, 1988, Prime Ministers Mulroney of Canada and Brian Bruntland of Norway called for a "Law of the Atmosphere" to reduce the risk of rapid climate change. In October 1988, the President of the Maldives galvanized global attention on the impact of the rise in sea levels, when he eloquently described the threat to his island nation in a speech to the UN General Assembly. In March 1989, Margaret Thatcher alerted the world to the political and economic risks of stratospheric ozone depletion. In June 1989, the Governing Council of the United Nations Environment Programme (UNEP) endorsed the idea of international negotiations leading to a draft convention on climate change. In July 1989, for the first time the leaders of the world's richest nations gave international environmental problems a high level of prominence in the discussions and in the final communiqu of the Summit of the Group of Seven (G7).

3. New and Emerging Dialogues Offer Continued Hope

The rising profile of environmental issues on the international agenda has provided many new opportunities for dialogue both between East and West, and between North and South. Although these issues have not yet penetrated the domain of traditional security debates, even here the context may be shifting to a broader focus.

In the near term, the most important fora for discussion of these issues may possibly be those with a predominantly regional character. For example, the bilateral discussions between the United States and Canada on clear air and acid rain have been a very useful platform for exploring the linkages between climate change, acid rain, and tropospheric pollution. The meeting on climate change held in Delhi, India, in February 1989, exposed a large number of senior analysts on the Indian sub-continent to the risks of climate change. This meeting explored some of the connections between the expanding coal consumption programmes, underway in both India and Pakistan, and the risks to agriculture, water resources and coastlines of a rapid climate change. In June 1989 a broadly conceived meeting at the East-West Centre explored the implications of these linked atmospheric problems for the nations of the Pacific Rim.

The July 1989 meeting of the G7 Summit in Paris dealt principally with the problems of the advanced industrial economies of Western Europe, Japan and the United States. Nonetheless, some effort was made to see environmental problems affecting these countries in a global context. The final communiqué from this Summit addressed the issue of not just what the G7 could do at home, but how their development assistance efforts could be reshaped to minimize the global environmental consequences of energy supply and use.

Once the understanding of the national impacts of these issues has reached a minimal level, the global dialogue can be successfully increased. Several fora now exist in which the linked concerns of environmental protection, economic growth and global security could be extended and enhanced. Two examples are outlined below.

On the ozone depletion issue, scientific understanding is now well advanced. Work on the manufacture of safer, alternative compounds, is approaching fruition for many applications. The follow-up discussions to the Montreal Protocol on Substances that Deplete the Ozone Layer are taking place in 1989 in the Helsinki round, where additional controls are being negotiated. The key challenge in this area is to expand participation in the Protocol's control regimes and to phase out the most dangerous compounds in the shortest possible time. The central issue is how to protect the rights of developing countries and their access to inexpensive substitute compounds, while minimizing the risks of further damage to the global atmosphere.

On the problem of rapid climate change, we are at a less mature stage. Initial discussions are in progress under the aegis of the Intergovernmental Panel on Climate Change (IPCC). Approximately 40 countries now participate in the three Working Groups of the IPCC. The Working Groups are scheduled to deliver interim reports in mid-1990. There are several fundamental weaknesses in the IPCC process. The Panel, commissioned in 1988 by UNEP and the World Meteorological Organization, is an ad hoc group with no formal charter. It has no mandate after 1990. The Panel has no mechanism to continue discussions or negotiations after its scientific assessment is completed in 1990. And it has no authority to mandate change in any policies or activities.

4. Institutional Issues and Challenges

The limited success of the international dialogue on ozone depletion and climate change, highlight some of the institutional issues and challenges which must be faced if global environmental issues are to be successfully addressed. One of the key questions is where and how to hold the discussions. While *ad hoc* meetings proliferate at an exponential rate, no permanent mechanism has been established to continue the dialogue. A number of suggestions have recently been made for the reshaping of existing institutions and the introduction of new institutions.

During the Pugwash Conference in Moscow in 1988, the Soviet Deputy Foreign Minister Vladimir Petrovski suggested that the UN Trusteeship Council might be an appropriate vehicle for such discussions. Later in the same year, Foreign Minister Eduard Shevernadze of the Soviet Union made this proposal in the UN General Assembly, noting that the original mission of the Trusteeship Council, to lead the remaining colonies on the path to independence, had now been largely fulfilled. Schevernadze noted that this dormant body might easily be revitalized as a Trusteeship Council for the planet.

In April 1989 Sir Crispin Tickell, the British Ambassador to the UN, made an alternative suggestion. He proposed that issues of global environmental degradation could best be aired in periodic special sessions of the UN Security Council. As many now recognize the security implications of environmental problems, the Security Council, Tickell argued, might be the most appropriate place to hold these discussions. Unhappy with holding these debates within the limited membership of the Security Council, others have suggested alternative frameworks within the UN system. One alternative involves expanding the mandate of UNEP to give it full status as an agency of the UN and giving it broader authority over these issues. Another alternative is to merge UNEP with the UN Development Programme and let the new agency deal with both environment and development.

Regardless of where and how the discussions are held, several knotty issues will remain to be resolved. The most important among these are issues of equity and responsibility. Before any international control regime can be implemented, agreement will have to be reached on the assessment of responsibility for past emissions and on the allocation of opportunities to continue releasing pollutants into the global commons. Should future "pollution rights" be based on population, on Gross National Product, on the history of past emissions, or on some other measure? Should pollution permits be bought and sold, like sowbellies or corn futures contracts? Should the costs of mitigation and adaptation be shared in some agreed proportion, or should each country bear whatever impacts "Mother Nature" sends its way?

Equally important to the resolution of these issues will be the role that the industrialized countries take in the transfer of new and less-polluting technology to the Third World. Unless the advanced industrial countries are willing to transfer these new technologies on concessionary terms, there is little incentive for the poorer countries to place a high priority on minimizing risks whose worst impacts may not be realized for decades. How to mediate the transfer of these technologies in an efficient and equitable fashion is an area that has received little serious attention.

5. Implications for Pugwash

With the 1988 Declaration of Dagomys, Pugwash entered the debate on environment and security issues. By offering a quiet and secure forum for dialogue between serious analysts from all camps, Pugwash could play a pivotal role in the coming decade. With their broad experience in negotiations, Pugwash members could help to untangle some of the knotty conceptual problems that have restrained national governments and limited progress in the past on these urgent international issues. Having focused for decades on measures to limit the causes of other risks whose impacts might not be felt for decades, if ever, Pugwash is in a unique position to support environmental negotiations which are necessarily entwined with issues of persistent uncertainty. Will or can Pugwash accept this challenge to broaden its traditional mandate? The Declaration of Dagomys is a gleaming symbol of hope. But the outcome will only be determined by those who actively join the struggle to promote economic growth in the context of environmental protection.

Development of Energy in China and Its Influence on the Environment

Zhu Jiaheng and He Gulian

1. Introduction

Energy consumption is increasing substantially as the world economy is rapidly developing. The influence of energy development on the environment is becoming more conspicuous; in particular, the pollution caused by burning fossil fuels is drawing ever greater attention from all nations.

Some experts hold that since the early 1980s, a second environmental upsurge has followed from the recognition of the magnitude of environmental pollution and large scale damage to the ecology balance. The first environmental upsurge occurred in the 1950s and 1960s. At that time, environmental pollution in industrialized countries was so serious that it threatened the life and security of the people, and exerted a tremendous influence on economic development. The United Nations "Conference of Human Environment", held in Stockholm in 1972, came just in time. It was a milestone for mankind: it recognized the importance of, and the need to make efforts for, environmental improvement.

The second environmental upsurge in recent years, shows more concern about problems which have an extremely harmful impact on large areas, including, for example, acid rain, ozonosphere destruction, and greenhouse effects. These problems attract a great deal of attention from countries all over the world.

The concerned authority in China has also begun to attach importance to these problems, especially in energy consumption. In China, the main fuels in cities and large industrial bases are coal and other fossil fuels, while in the countryside 75 per cent of energy is dependent on biomass energy; this pattern undoubtedly exerts its influence on environment pollution and undermines the ecology balance. In recent years, China has started studying the future programme of energy development, investigating the necessity of reforming the energy structure, and strengthening research on environmental protection, monitoring and cleaning measures. It has promoted international exchanges and cooperation, and tried its best to draw a reasonable blueprint for the solution to the all-round energy and environmental problems.

2. The Development and Characteristics of China's Energy Structure

Since the People's Republic of China was founded in 1949, energy production has increased rapidly. In 1980, the national gross output of industry and agriculture was 16.2 times that in 1949. Energy consumption was 25.7 times larger. During the period 1949–1980, the average growth rate of the national gross output of industry and agriculture was 9.4 per cent, while that of the energy consumption was 11 per cent. In 1987, the gross output of primary energy sources in China reached 910 million tonnes in terms of standard coal. Among these sources the output of raw coal amounted to 920 million tonnes, occupying the first place in the world. The production of crude oil was 134 million tonnes, occupying fifth place; electric energy production reached 497.3 terawatt-hours (TWh), occupying fourth place; hydraulic electric energy production was 100.2 TWh, occupying fifth place; natural gas production was 13.5 billion cubic metres.

In the year 2000, the planned goal of China's energy production is: 1.4 terawatt-year (TWy) of primary energy, gross electric energy production 1200 TWh, with 240 GW installed capacity.

An intermediate-to-long-term economical and annual energy forecast demand of China is given in Table 1 (from the 1st issue of the Chinese journal "Nuclear Sciences and Engineering" in 1988).

Table 1. A Forecast of China's Intermediate-to-Long-Term Economical and Energy Demand

Target	Unit	Year				
		1980	1985	1990	2000	2030
Gross National Product	10^8US$	2736	4597	6600	12030	51990
Annual Energy Demand	TWY	0.602	0.841	0.991	1.414	3.45
Annual Electric Energy Demand	TWh	300.6	407.3	550.0	1200	5186
Energy Consumption of Output Value	TWy / $10^{12}$$	2.20	1.83	1.50	1.18	0.66
Proportion of electric energy in the Gross Output of Primary Energy	per cent	20.6	19.3	21.1	30.5	49.6
GNP/Capita	$ / person year	277	439	593	962	3710
Annual Energy Consumption Per Capita	10^{-9} TWy / person year	0.61	0.8	0.89	1.13	2.46
Annual Consumption of Electric Energy Per Capita	kWh / person year	304	389	494	960	3700

Notes – Output value is calculated according to the prices in 1980
– At the time of forecast, 1 US$ = 1.55 yuan Renminbi.
– The figures in 1990 are those envisaged in "the Seventh Five-Year-Plan".

From Table 1, it is clear that the energy demand in the near 2000 will be double that in 1980. In 2030, it will be nearer six times that. As to the energy consumption per capita, it is very low in our country. It was only 0.8 kWy in 1985, one quarter of the world average level. In order to make the living standard of Chinese people reach the average level of middle-developed countries, the energy consumption per capita will have to be about 5 kWy, and the gross energy demand will reach more than 6 TWy. It is a very hard task.

China is quite rich in coal, petroleum, natural gas, hydro-energy, uranium mineral and other energy resources. The supply of various energy sources, therefore, can be wholly internal. Especially for coal; China has large reserves and good assortment, allowing coal to serve as the mainstay of her energy sources. At present, it makes up about 75 per cent of the total energy consumption. The fuel for 78 per cent of thermal power is coal. The January 1987 issue of "Energy Statistics Monthly" in the USA, contains a "Primary Energy Consumption Constitution in World Major Countries in 1986" (Table 2). As this Table shows, one of the prominent features in China, compared with other countries, is that coal is its main fuel. The proportion of coal in China is 2.5 times the world average.

Table 2. Primary Energy Consumption in World Major Countries in 1986

	the world	China	USA	Japan	FRG	France	UK	Italy
Consumption (TWY)	10.81	0.8088	2.659	0.5666	0.3783	0.281	0.2991	0.1902
Consumption Constitution (per cent)								
Petroleum	38.0	17.1	43.2	55.8	42.5	43.0	36.2	60.9
Natural Gas	20.0	2.3	22.4	10.2	15.5	12.5	23.5	21.7
Coal	30.0	76.0	23.3	18.8	30.1	10.0	32.4	9.8
Nuclear Power	5.0		6.1	10.3	10.9	32.8	7.5	1.8
Hydraulic Power etc	7.0	4.6	5.0	4.9	1.0	1.7	0.4	5.8
Total	100.0	100.0	100.0	100.0	100.0	100.0	100.0	100.0

3. Greater Negative Environmental Impact of Coal-Fuelled Energy Sources

In general, the worldwide influence of energy consumption on the environment includes the following areas: emission of waste heat and concentration of carbon dioxide may affect the climate of the earth; carbon dioxide and heavy metals cause the pollution of the environment; petroleum products cause the pollution of the oceans in transportation across the seas; and accumulation of the isotope krypton-85 contaminates the atmosphere.

The most important influence of energy consumption on the regional environment is injury to the human body, caused by the emission of chemical substances. The emission of waste heat also undermines the ecology of the water systems.

Coal, as well as other fossil fuels, are energy sources causing serious pollution to the environment. An average thermal power plant of 1 GW drains off 1 million tonnes of soot, more than 60 thousand tonnes of sulphur dioxide, and 630 kg of the strong carcinogen benzopyrene, per year. 73 per cent of floating dust and 90 per cent of sulphur dioxide, drained into the atmosphere, come from burning coal.

The sulphur dioxide and nitrogen oxide, poured into the atmosphere can be turned into substances rapidly solvable by the cloud drops, becoming sulphuric acid and nitric acid through chemical reactions. This leads to acid rain. The concentration of carbon dioxide and other gases in the atmosphere results in the absorption and re-radiation of some longer-wave infra-red radiation, causing the temperature to increase on the earth's surface, This, the so-called "greenhouse effect", has an increasing trend. According to research findings of Professor Zhu Kekhen, former Vice-President of the Academy of Sciences in China, the change of air temperature in the past 5000 years in China is about 2°C. But at the Conference convened in Austria in 1985, by the World Meteorological Organization, the World Scientist League and the United Nations Environmental Bureau, it was stated that, by the thirties of next century, the density of carbon dioxide in the atmosphere will be doubled, so that the world air temperature will increase 1.5 to 4.5°C in fifty years. In such a short period of time, the enormous increase of air temperature will melt part of the icebergs in the Antarctic and Arctic, causing an increase in the sea level. The increase of air temperature will also produce more sterile areas and migration of the rain fields, increase natural calamities, and decrease the grain yield, bringing great suffering to mankind.

In our country, owing to the very large population, big gross energy consumption, and sole reliance on coal, atmospheric pollution is very serious. Monitoring data on certain cities in China tells us that in 89.4 per cent of the cities investigated, the density of particulate matter in the atmosphere exceeds the state standard. In cities like Chongqing, Guiyang, Changsha, Nanchang, etc., acid rain pollution was observed and it appears that it is becoming more widespread. Among the forty cities monitored by the worldwide environmental monitoring system during 1980–1984, the five cities monitored in our country – Shengyang, Xian, Beijing, Shanghai and Guangzhou ranked second, third, fifth, ninth and tenth respectively in the density of particulate matter (Kuwait in the Middle East ranked first in this survey). In the city of Benxi of Liaoning province, the visibility is only 30 to 50 metres during nearly half of the year due to atmospheric pollution. Our country has become one of the most seriously polluted countries in the world.

4. Comprehensive Arrangement of Energy Sources and Environmental Protection

The energy industry is an industry with a large investment and a long cycle in capital construction. In developing energy sources we must consider not only the demands of economic development but also the demand of decreasing environmental pollution. It is necessary to have long-term, rational, overall strategies.

The disadvantages of China's energy structure, heavily reliant on coal, are obvious. They include shortage of transportation, serious pollution, and the profound and lasting influence on the earth's climate. So it is necessary to plan the reform of the energy structure according to our present energy resources before it is too late. A comparatively rational plan is to transform the present energy structure gradually into a pluralized structure, with coal, water and nuclear as its three pillars, and other energy sources as its supplements.

Hydro energy: waterpower resource is an energy source which is continually regenerating, along with the periodic return of hydrographical circulation. It is one of the cleanest energy sources among the regular primary sources. Our hydro energy is one of the richest natural resources in the world. The theoretical reserve throughout the country is 680 GW of which 378 GW may be exploited. Up to the end of 1987, the installed capacity of various hydroelectric stations was about 30 GW and that under construction was near 20 GW. In addition, there are more than 300 GW to be exploited. There is considerable potential in this field.

In order to meet the energy demand and decrease reliance on coal as soon as possible, we should adopt a policy of speeding up the development of medium-sized hydroelectric stations. The investment per kW in a medium-sized hydroelectric station is about 20 per cent higher than that required for a large scale one, but the time required for going into operation is about 50 per cent shorter. It has the advantage of moderate scale, short construction time and quick results.

Nuclear energy: China is a country rich in uranium resources. The proven uranium deposits are enough to make a number of pressurized water reactor stations of 15 GW capacity operate for many years. Moreover, if we begin to develop breeder reactors in time, the utilization ratio of these resources can be raised sixty or seventy times and the nuclear energy resources will be enough for use even for hundreds of years. Of course, if in future nuclear fusion power stations could be developed, the nuclear energy resources will be infinite.

Nuclear power is an energy source which is safe, economic, and less harmful to the environment. In spite of the fact that people have been doubtful about the development of nuclear power after the Three-mile Island and Chernobyl accidents, on the whole, most nuclear power plants in the world have continuously been in safe operation. Nuclear security may be further improved through better co-ordination by technical means and operating methods.

Our country's construction of nuclear power has already started. The pressurized-water reactor station of 300 MW in Qinsham, which was designed, developed and built with our own resources, will be completed to generate electricity in 1990. The electric power of the imported nuclear power station at Daya Bay in Guangdong is 2 × 900 MW estimated to generate electricity in the grid net in 1992–1993. Besides, the nuclear power plant of Qinsham has been preparing in advance for the second phase of completing two 600 MW generating units. According to the present preliminary programme, the gross installed capacity of nuclear power in China will be 3.3 GW in the year 2000, and about 30 GW in 2015.

At present the guiding principle of developing nuclear power in China has been laid down, namely "relying on our own efforts while seeking cooperation with outside". The pressurized water reactor of 600 MW has been chosen as its main reactor, so as to accelerate our nuclear power development.

Most Chinese and foreign experts hold that nuclear power is the most realistic energy source with a bright prospect. To develop nuclear energy on a large scale is one of the fundamental measures for solving energy problems in our country, as well as in the whole world, by the end of this century and the early years of the next century.

In addition, we should pay great attention to the development and utilization of solar energy, wind power and other clean energy sources.

5. Strengthening International Exchanges and Cooperation

The upsurge of the environmental issue involves a wide range of problems, including vast polluted areas and complex pollutants. Some pollution comes from economic activities, other from everyday life. Some comes from economically developed countries, other from developing countries. So it is necessary to develop more cooperation throughout the world. We suggest the development of international exchanges and cooperation in the following four areas:

1. Strengthen exchanges and cooperation in environmental monitoring, and management.
   The environmental monitoring of various countries is an important means to get a clear view of the trend of environmental pollution. Strengthening exchange and cooperation will deepen the monitoring work throughout the world.
2. Strengthen exchanges and cooperation in measures to clean the environment. While strengthening the study of acid rain, and undermining of the ozonosphere, and greenhouse effects, we should vigorously develop the study of measures to clean the environment, and popularize those with economical and effective results. Developed countries should transfer modern, low-pollution techniques to the Third World.
3. Strengthen exchanges and cooperation with regard to nuclear power security. Any nuclear accident occurring in the world affects not only the environment

and population in one's own country but other countries as well. So the nuclear power security problem has become the central issue of deep concern to the world nuclear industrial circles. It is necessary to strengthen exchanges and cooperation in this field. As the nuclear power undertaking in our country has just begun, it is especially important to draw lessons from all previous accidents abroad. In international exchanges and cooperation, we should do all we can to reduce or eliminate possibilities of accidents with their potential serious results, and to make our nuclear power plants safe and in reliable operation.

4. Strengthen exchanges and cooperation in developing technology of clean energy sources.
   Solar energy, wind power, etc. are clean energy sources of great potential. But it is necessary to solve the technical problems of development and utilization. If we can make good technical progress in this field by effective international cooperation, it will bring tremendous benefit to the people of China and the whole world as well.

Part VIII

East-West and North-South Cooperation

Trade and Finance: The Bitter Truth

Anatoly Gromyko

1. Introduction

The situation in the 1980s has noticeably changed the tone of the discussions regarding the possibilities of an effective solution to one of the most poignant problems of modern times: the under-development of the Third World countries. The sharp drop in rates of economic growth throughout those countries, especially in Africa, has brought more than just new disappointments. The 1980s have continued the period during which negative rates of growth were registered in Africa. According to estimates of the Economic Commission for Africa (ECA), the living standard of an average African, say, in 1988 amounted to just 80 per cent of the 1980 level. An opinion seems to be shaping among specialists in international affairs and public figures that for an overwhelming majority of the newly-free states the former dynamic rates of economic growth are now unrealistic.

Optimism or pessimism with respect to the future of the Third World is a matter of faith. It should be noted, however, that the grave consequences of the protracted crisis in various parts of that region, not only threaten the economic security of many countries which are mostly at the medium and bottom tiers of development, but also threaten a serious destabilization of the entire spectrum of international relations. This was openly stated at the Thirteenth Special Session of the United Nations, dealing with the critical situation in Africa. That is why I think the time is ripe for coming up with new, large-scale international initiatives, so as to reinforce, on the basis of a balance of mutual interests and recognition of the special needs of the under-developed countries, their own efforts with truly effective economic assistance. Without such initiatives, the Third World is condemned to stagnation and even collapse, although some may find this term too strong.

2. International Trade: A Change in the Balance of Interdependence

The deterioration of the financial and economic situation in developing countries since the start of the 1980s was caused by a series of factors. To a large extent, it resulted from a near-sighted and unbalanced economic policy, an unbearable burden of non-productive expenses, and a low and often negative payoff of capital investments whose expediency was quite doubtful. However, in my opinion, the beginning of this sequence of crises resulted in a considerable deterioration in the position of the developing countries in world trade, dramatic growth in their foreign debt, and a noticeable decline in the inflow of external resources.

Probably the most important among these three major factors, at least for the countries with a hypertrophied economic structure with respect to agricultural raw materials, is the reduction of their participation in the new structure of the international division of labour, under the impact of the scientific and technological revolution.

There are no more doubts today that the decline in the rates of growth in vast areas of the Third World was a reflection, not only of the protracted global crisis of 1980 to 1982 and of a drop in world prices for raw materials, but also of some more fundamental changes. The latter strongly influenced the balance of the former asymmetrical interdependence of the two sub-systems of the world capitalist economy.

Specialization in raw materials proved to be an unreliable foundation. The predicted strength of the oil cartel turned out to be illusory, not to mention the other associations of the countries producing various raw materials. In contrast to this, the economy of the industrial centres demonstrated a high degree of adaptability to the oil shocks. The acceleration of the trend towards raw material conservation, which first appeared in the West and Japan already in the 1970s, the use of waste-free technologies and synthetic materials, and the adoption of effective measures to strengthen the national raw material and food supply sectors of the economy, appreciably reduced the material- and energy-intensity of production, and precipitated a sharp drop in the global demand for many and especially the traditional raw materials and fuel. This seriously affected the export of agricultural raw materials and food from tropical countries, notably as a result of a drop in the elasticity of the growth of the demand for it in relation to the growth of incomes in advanced countries.

In my opinion, all this compounded the negative trends for the Third World, which were, furthermore, accompanied by a steady rise in the prices of industrial goods (largely under the influence of transnational corporations). Most developing countries quickly lost their foreign-exchange reserves, and this naturally resulted in grave consequences for internal accumulation and growth. This negative situation was noticeably exacerbated by a narrowing – mostly under the impact of automation in the industrialized countries – of the sphere of use in the Third World of the comparative advantages associated with the lower labour costs. Alongside other factors, this substantially undermined the competitiveness of the commodities produced on the periphery of the world economy.

There is another factor that should be taken into consideration here. The structural re-building in the industrial centres is being accompanied by renewed manifestations of protectionism which are extremely painful for the Third World. According to the estimates of the International Monetary Fund (IMF) and of the International Bank for Reconstruction and Development (IBRD), the losses of the developing countries resulting from all forms of protectionism are approximately twice as large as the inflow of resources through the official aid channels.

Finally, the developing countries' trade is negatively affected by the wide fluctuations in foreign exchange rates. The paradox of the current stage of the internationalization of the world economy is that this trend considerably undermines the former balance of interdependence of the two subsystems (the central and the peripheral) of the world capitalist economy. In effect, vast areas of the Third World are being pushed off the main road of global development.

Needless to say, this negative trend manifests itself differently within the context of economic differentiation in the Third World. Nor can we rule out the possibility of a closer involvement of particular sectors of the contemporary economies of developing countries in the ongoing transformation of the international division of labour. Nevertheless, even in the countries which have advanced a long way and are close to becoming new industrialized countries, such adaptation is strongly complicated by the expansion of the structural gap in the dynamics of the development of productive forces at the centres of the global economy (the transition onto the track of a post-industrial economy) and on its periphery.

The worst affected countries are those at the bottom of the peripheral pyramid. This can be best illustrated by Africa south of the Sahara, which suffered great damage from the catastrophic destabilization of its trade with the rest of the world. For example, as a result of the deterioration of the terms of trade alone, it annually lost (excluding Nigeria) about $3 billion between 1979/1981 and 1985/1987. One should add to this the damage from the strengthening in the 1980s of the tariff barriers and quantitative restrictions on the export of cotton, textiles, sugar, tobacco, vegetables, fish, cocoa and semi-finished products from mineral raw materials.

The economic and social conditions in African countries, just as in the other backward parts of the Third World, are based on the conservatively traditional or hybrid forms of production. All this is naturally fraught with a further reduction of those countries' participation in world trade and economic exchanges and, consequently, with a reduction of their development potential. Suffice it to cite the following remarkable facts: the share of Africa (excluding South Africa) in world exports shrunk form 4.7 to 2.1 per cent between 1980 and 1987, while per capita imports dropped down to the level of twenty years ago.

The growing inability of most developing countries to adapt more or less effectively to the qualitative changes in the international division of labour, objectively strengthens the need for the regulation of the mechanisms of their structural adaptation to the prevailing trends in the world economy. These days, such regulation is effected mostly through the so-called programmes of structural adjustment worked out under the auspices of the IMF and the World Bank. These programmes are based on neo-liberal doctrines which are aimed at stimulating the export potential of the former colonies and semi-colonies. The main emphasis is placed on the use of price regulators: devaluations, elimination of imbalances in interest rates, and easing of customs regulations. There are also plans for raising the purchasing prices and for reducing the subsidies to farmers. Without going into the details of this matter, I would like to note that the largely justified use of the market instruments for the regulation of trade still cannot by itself, without more effective levers of foreign economic policy, appreciably influence on a long-term basis the dynamics of the structural changes in the world economy, which scarcely depend on developing countries, or are even totally beyond their control.

That is why it is so important to find new points of contact to define the properly understood interests of all countries of the international community, in order to lay a durable foundation for the development of a programme of cooperation that would envisage closer regard for the needs of the developing countries for the expansion of their trade. In my opinion, bold steps should be undertaken within the framework of which an approach for a global revision in the General Agreement on Tariffs and Trade (GATT) of the entire spectrum of protectionist policies, including the lifting of quotas and of limitations on exports from African countries and other under-developed Third World nations, and the expansion of the preferential conditions for them (after the pattern of the Lomé Convention), not only in the sphere of the export of raw materials, but also of industrial goods.

It would be also proper to speed up the codification of the procedure associated with the establishment of the United Nations Conference on Trade and Development (UNCTAD) Common Fund for Commodities, and to seek out new opportunities for replenishing the funds for compensatory financing of foreign-exchange losses arising out of *force-majeure* drops in market demand for the raw materials of the least developed and other newly-free states worst affected by the crisis. Parallel with this, it would be very important to work out, on a collective basis, target-oriented programmes of assistance with respect to the diversification of the export structure of the countries specializing in a single commodity, and to adopt a system of measures facilitating the adaptation of their trade to the prevailing trends in the world economy. And, last but not least, any such regulation should include stimulation of the mechanisms of trade and economic cooperation, on the basis of the expansion of joint forms of activities in the sphere of private enterprise, notably through collective provision of credit for projects enhancing the potential of mutual trade.

In other words, a well-thought-out and flexible approach to trade relations as the prime mover of development, would meet the interests of not only the Third

World countries but of the entire international community as well. Destabilization of the exports of such vast regions as Africa undermines their solvency and seriously complicates the sales of goods from the advanced countries, and this, for its own part, hampers the development of the world economy.

3. Debts and External Financing: A Sphere for Joint Action?

The debt crisis has affected an overwhelming majority of developing countries of which more than 60 have lost their solvency and many others are on the brink of a financial abyss. The hardest hit by the debt crisis is Africa to which I will try to give closer attention in this part of my paper. Africa's debt currently stands at 230 billion dollars. This amounts to 70 per cent of the combined gross domestic product of the African states. The payments to service this debt will apparently exceed 30 billion dollars in 1989. Even given the rescheduling of debts, these payments claim about a quarter of the African states' foreign-exchange earnings. Even the intensive economic growth in the industrialized Western countries, which has been a subject for self-satisfaction on the part of their political leaders in recent years, has failed to ease the continent's debt burden in any appreciable way. The said growth has not affected Africa.

Just as with all the other problems arousing the concern of the international community, the hopes for the future with respect to this one are associated with the joint efforts of all states: both debtors and creditors. As the debt crisis grows sharper, the isolationist moods which gained currency in Africa over the previous decade have started to fade. Without additional benefits to the continent from the creditors the ever-mounting difficulties cannot be overcome. All the major economic initiatives of Africa underscore the importance of constructive cooperation of all groups of countries in the settlement of the pressing problems of the continent. It is also obvious that the measures which have been undertaken by the West are totally insufficient for resolving the debt problem. The situation in Africa has clearly got out of control and hence urgent measures are needed to overcome the crisis. The main factors of the debt crisis are beyond the control of most debtors. For the least developed nations the only reasonable solution would be a substantial reduction of the debt burden.

Two-and-a-half years ago a special session of the UN General Assembly approved an emergency Programme of Economic Revival of Africa. It was intended for a period of five years, but today, in the middle of its implementation, we can already sum up some results. The most crucial principle of this programme was reciprocity of the commitments undertaken by Africa and by the international community. Sharing of responsibility means sharing of the burden of the crisis as well. No specialist today could insist that Africa is undertaking insufficient efforts in that direction. The former economic policy of many African debtor nations was unrealistic and ineffective. Over the past three years, however, resolute measures have been taken in most countries of the region to enforce budgetary discipline, to close hundreds of loss-making enterprises in the public sector, and to dismantle the bureaucratic structures which blocked economic enterprise. All African countries are liberalizing their currency policy. The exchange rate of most African countries is being adjusted to their real purchasing capacity. This means substantial devaluation and a hope for the expansion of exports. The African governments are also correcting the entire system of domestic pricing which used to hamper economic activity for decades. In particular, it deprived food producers of the incentives to work for the domestic market, and created the dependence of Africa on the import of grain, sugar and other foodstuffs.

Why then all that anxiety in the latest reviews of the African economy? Experts understand what is clear to any man in the street: financial and economic stabilization, which inevitably involves further "tightening of belts", has definite and, in the African context, quite narrow limits. The efforts at stabilization undertaken by Africa are certainly needed (this is not disputed by Africans themselves), but their positive effect will not manifest itself at once, and there will be an inevitable lag of three to five, or even more years. But the social and political consequences of the current reforms will be felt right now, by everyone in his own pocket. Structural stabilization is a painful process even in industrialized countries. Everyone remembers the sharp problems which such stabilization created, say, in the UK at the end of the 1970s and the beginning of the 1980s. Given the extremely low level of incomes, the social structures of African states, subjected to such a severe trial of strength, may simply collapse under the weight of the problems born of stabilization. This explains the ever louder voices of those who are concerned over "adjustment fatigue". That is why a recent UN report stresses that the measures of the international community are insufficient, and that unless additional efforts are made, the stabilization may grind to a halt and the vicious circle of economic stagnation will not be broken.

The African continent needs a long-term alternative capable of laying a healthy foundation for economic progress. The initiatives of Africa are aimed at the development of exactly this kind of an alternative strategy. The West, too, seems to be moving towards the recognition of the inevitability of concessional solutions. The latest examples of this recognition are the package of French proposals and the US "Brady Plan". Despite the absence of details, this plan has been welcomed by a wide range of debtor nations, from Brazil to Zambia. I think that this plan is oriented to the states which are comparatively mature economically, hence the majority of African states will gain little from it. What is important, however, is the principle built into the plan: for the first time it recognizes that it is not enough to broaden external financing and that it is also necessary to reduce the already accumulated debt.

The shift in the policy of the Western creditors towards realism can substantially benefit Tropical Africa, in particular. In the first five years of the debt crisis, most Western initiatives were addressed to the biggest debtors whose problems threatened the stability of the international credit system. In contrast to this, in the past two years the industrialized nations have at last addressed their efforts to the restoration of the solvency of the most impoverished African states. It should also be noted that all the latest initiatives of the West are consonant with the debt plans of the Organization of African Unity (OAU) and the ECA. Initially, those plans received a cool response from the West, but today many of their elements are identifiable in the decisions of the IMF, the IBRD and other creditors. It appears that the proposals of the OAU and the ECA are actually working for Africa's benefit as instruments of collective pressure on the creditors in the direction of realistic solutions.

For its own part, the Soviet Union, which realizes that the debt is today the main obstacle on the way of Africa's economic progress, has stepped up the search for solutions to this extremely poignant problem. In principle, the debt to the Soviet Union amounts to no more than 15 per cent of the African continent's overall debt. Most credit agreements between the Soviet Union and African states envisage the possibility of the repayment of debts not in hard currency but in ways less burdensome for the debtors, for example, with deliveries to the Soviet Union of traditional export commodities. In the first years of the debt crisis, this fueled unfounded optimism regarding the prospects of servicing the debt to the Soviet Union: it was hoped that the crisis would pass the Soviet Union by. The reality, however, proved much harsher. The crisis forced the African states to take all conceivable measures to increase their foreign-exchange earnings. Some of them rechannelled to other markets that part of their exports which was formerly set aside as payment of debts to the Soviet Union. Besides, the traditional Soviet approach to the debt problem was based on the fact that without being a member of the main institutions regulating the movement of capital (the IMF, the IBRD, the Paris Club) the Soviet Union could not bear responsibility for the development of the crisis. Without being a member of the leading international financial institutions, the Soviet Union could not influence the processes substantially affecting the course of its cooperation with African states. It is indicative that some of the debtor nations approached the Soviet Union with the request to take part in the discussion of their debt problems within the framework of the Paris Club in the hope that this would help work out more flexible and concessional terms of debt rescheduling. Who knows, if that had happened in the early 1980s, the debt problems of Africa might have been eased.

In the past four years, since the start of *Perestroika* in the Soviet Union, the debt problem has been consistently rising in the list of priorities of Soviet foreign economic policy. In its fullest form, the Soviet approach was laid down in the address by Mikhail Gorbachev to the 43rd session of the UN General Assembly. The Soviet debt initiative is primarily addressed to the poorest countries of Africa. The confrontational moods which appeared in the debt sphere in the first years of the crisis have done no-one any good. That is why the Soviet initiative proceeds from the desirability of reinforcing those positive elements which took shape in the course of the concrete discussions aiming at resolving the debt problems of the developing nations, and at strengthening the already existing international organizational structures.

In today's world many things are interdependent, and the decisions of states in the economic sphere can hardly be separated from their approaches to foreign policy. The integrity of the approach to foreign policy is a major distinction in the new political thinking in the Soviet Union. This explains the special importance of its drive to bring the discussion of the debt problem to a conference sponsored by the United Nations, which has the mechanisms for working out a consensus. Consensus is necessary because separate solutions lead to deadlock. In my opinion, the nature of the global problems is such that there are often no direct culprits for them on whom the blame can be pinned for the development of a crisis situation. This is exactly how it is with the debts also. If the debt crisis has taught us anything at all, it is the futility of mutual accusations. This is why the former Western strategy of resolving the crisis, based on its explanation which stressed "the mistakes and irresponsibility of the debtors", turned out to be such a failure. It led to too harsh terms being imposed on the debtors, and was thus doomed to failure.

I think that the way out of the debt crisis consists in the rapid establishment of a system of international economic security and in an energetic restructuring of international economic relations. Here I foresee the following priority steps: reduction or long rescheduling of payments under foreign debts depending on the economic development figures; democratization of international trade and the removal of protectionist barriers; renunciation of additional interest charges in rescheduling debts; reduction of interest exchange under bank credits; stabilization of national currency rates; and government support for market mechanisms for settling the debt of the Third World, including the establishment of a specialized international agency for buying debts at discount. As for the least developed countries of the Third World, there should be a long – up to a hundred-year – moratorium on the repayment of their debts which in many cases should be written off completely.

Since the measures for the international co-ordination of efforts to stabilize the solvency of African states can take many years, it would be advisable to carry out in the near future a wholesale rescheduling of Africa's debt to the Soviet Union on a unilateral basis. It is imperative to write off the official debt of the 18 least developed countries of the continent where annual per capita incomes are below 500 dollars. There is also a need for a search for flexible and concessional mechanisms of debt rescheduling for more advanced countries. Such a debt initiative will demonstrate foresight and realism which are so badly needed in resolving global problems.

The elements of realism appearing in the debt strategy of the West, and the active policy of the debtor nations themselves which have come up with a series of well-thought-out, comprehensive proposals, inspire the hope that the collective search for solutions in this sphere will produce the desired effect.

4. Conclusion

In conclusion, I would like to point out that the influence of the external factors on the dynamics of relations with the Third World nations cannot be reduced to merely quantitative parameters. The internalization of international affairs, the growing interdependence of national economies, and their gradual involvement in the international division of labour, all raise the degree of dependence of their development on structural changes in the world economy. Looking back today, one can speak of certain weaknesses in the programme for a new international economic order and of its unjustified maximalism. At the same time, one cannot overlook the fact that, on the whole, it is aimed at right and historically achievable goals. This refers to the establishment in international relations of democratic principles based on the unquestionable respect for the sovereign rights of all states, removal of discrimination, artificial barriers and group egotism from international trade, and collective responsibility and solidarity in the face of the global problems which hold back progress.

A further policy of disarmament and the establishment of a reasonable foundation for international security, notably in the economic sphere, can lay a firm groundwork for resolving the problems of underdevelopment.

New Approaches to South-South Cooperation

Surendra Patel

1. Introduction

The importance of South-South Cooperation has been for long recognized, both at a general philosophical level and in formal resolutions on the subject. Several reasons have been advanced for its promotion: for instance, the expansion of trade through mutual reduction of tariff barriers, as was done through the establishment of the Common Market in Europe; the appropriateness of the development experience of more advanced Third World countries to other less advanced ones; the complementarity of output and trade structures of the developing countries; the importance of learning from each other's relevant experience; and the strengthening of the solidarity of the Third World, and the Non-Aligned countries.

2. The New Setting

These reasons have been a useful starting point. The past experience of regional and sub-regional trade and other forms of cooperation, however, have shown that successes achieved through conventional approaches have been marginal. Many of the promising schemes for such cooperation have either faded away or considerably weakened over time.

Moreover, several new factors have made their appearance on the world stage. They require a close re-examination of the lines pursued so far in the past. There has been a slowing down of the world economy; a breakdown of the Bretton Woods structures established at the end of the Second World War; and with the "Golden Age" of growth coming to an end, several developing countries are indeed facing very serious crises in their balances of payments, brought about by the decline in their export growth and the heavy burden of debt servicing and adjustment policies imposed upon them.

Together with these adverse phenomena, several promising developments have also emerged. There has been a reduction in Cold War tensions. Some positive moves have been made towards superpower understanding. The pace of disarmament negotiations has quickened. While most countries of Latin America and Africa are facing severe crises, several developing countries in Asia have continued their rapid growth. Some of them are very large, particularly China and India. Their sustained growth has opened up new possibilities of these dynamic countries acting as a vigorous engine of economic growth. Among the positive factors should also be listed the fall of dictatorships and military power in many countries of the South. A wave of democratization is indeed sweeping many countries of the South. It is opening up new avenues for active popular participation in shaping the destinies of the countries concerned. There are also welcome signs of reduction in long-festering sub-regional and regional armed conflicts.

Among these new factors, I should also add perhaps the simultaneous appearance of the slowing down of the world economy and the emergence of new technologies, such as electronics, informatics, bio-technology, new materials, super-conductivity and fusion. Such simultaneity of timing of these twin phenomena had precedents in the past, for example, in the "Long Depression" of 1874–1896, and in the "Great Depression" of the 1930s.

There is, however, an unfortunate air of despair around. Food scarcity, falling exports, heavy debt servicing, severe pressures on the balances of payments and mounting internal tensions, have all converged to pose grave problems for the developing countries. The pressures facing them are severe. They have to be assessed on the basis of a proper understanding of both the strengths and weaknesses of the South. The trade-offs between the short-term and the long-term pre-requisites of accelerating the development process need to be carefully weighed. And yet the concern with the contemporary crises should not be allowed to overshadow the real advances that the Third World has made since 1950. Its overall Gross Domestic Product (GDP) has increased six-fold and the per capita GDP has increased two-and-one-half times. Its industrial output is 11 times higher. The volume of gross capital formation now is 15 times larger than in 1950. Enrolment at the tertiary level in universities and institutions of higher learning has simply exploded, rising nearly 25-fold. The South of 1989 is thus vastly different from its crippled image of 1950.

The consequent structural changes in the Third World have been profound. The share of agricultural output in GDP has fallen from around 37 per cent in 1950 to only 17 per cent now. That of industry has correspondingly risen from about one-fifth to over one-third. Increasingly, the output structures in both the North and the South are beginning to look alike.

It is true, of course, these overall figures conceal the differences in development among countries. The countries of the South have not all shared equally in these advances. Some have done much better. A positive way of looking at this asymmetrical performance is perhaps to treat it as opening wide the window of opportunity for enlarged cooperation. Every crisis, as the Chinese say, is at the same time a danger and an opportunity. The current crisis too contains both the negative and positive aspects. The new strategy of development of the South should be forged taking into account both these factors.

Attention must also be given to the crises facing the developed countries, both capitalist and socialist alike. They too are experiencing many problems.

They are faced with high unemployment, slow growth, falling savings and investment levels, slow industrial development, heavy surpluses and deficits on their current and domestic accounts, intense trade rivalries, aging populations, domestic social tensions, massive environmental degradation, and so on. These negative phenomena in the developed countries coexist also with the decline in superpower tensions, real progress towards disarmament, and the opening up of new horizons for accelerated technological development. The South need not feel lonely in facing its own crises. The East and West too have their own special brand of problems.

3. Some Ideas on New Approaches for South-South Cooperation

I have only touched upon some of the new factors on the world scene. I hope they may serve as a frame of reference for evolving new approaches to future South-South, as well as North-South and East-West cooperation. Here is a subject which has long been under discussion. It is now being actively explored by the South Commission, headed by Mwalimu Nyrere, the former President of Tanzania. The ideas described below are, therefore, tentative and are advanced with a view to their more systematic consideration in the future.

3.1 Unilateral Initiatives by the South

There is now an impasse in the North-South dialogue. Proposals continue to be advanced to break it. But, unfortunately, none of them appears to have made any significant impact. This would seem to suggest that there is perhaps a need for a somehow different approach. A parallel may be drawn with East-West relations which were for so long frozen in the mould of the Cold War. A change of track in East-West discussions required some bold new unilateral approaches. The initiatives by the Soviet President, Mikhail Gorbachev, appear to have broken the ice. In a similar fashion, the Third World too could perhaps undertake several unilateral initiatives. The point to stress here is that such initiatives should not be conditional upon what the others do or do not do.

Up to now, the South has been knocking at the doors of the North asking for agreement to changes in existing international economic relations. It has advanced several proposals of considerable merit for reshaping existing relations. But the response from the North has been cold, if not negative. Now, if these proposals are good and reasonable by themselves, why cannot the South, as a whole or several countries within it, unilaterally start to put into practice those proposals which are relevant to fostering South-South cooperation?

Several ideas may be considered. They may not be found immediately acceptable. After all, the world is full of sceptics who always say: that just cannot be done. For such Mr Cannots, the South desperately needs persuasive Mr Cans. If a step is right, it is not sensible to wait for others to agree to it before some country or group of countries show the needed courage to make a start. Of course, the particular step has to be carefully chosen. There are at least three broad areas in which such unilateral steps could be considered:

Disarmament. The South has been pressing the North to move fast towards real and substantial disarmament; and to divert the resources saved to the development of the South. But at the same time, the South itself has been wasting enormous resources in military pursuits, including imports of defence equipment and technology. This competitive madness has to be stopped somewhere. Besides saving resources, the South would then be able to reduce tensions among neighbours and to settle long-festering regional conflicts.

Why cannot the South, as a whole or some countries within it, unilaterally call a halt to this unpardonable waste of precious resources? It can then start diverting these resources to productive uses. This can help raise the volume of capital formation and release valuable foreign exchange resources for developmental imports. It would reduce external involvement in intra-South conflicts.

New winds of peace and progress have begun to sweep away the foul debris of the Cold War, and the conflicts it generated, supported and sustained. The South must show that it too is with the wind. Its credibility will be impaired if it lags behind or falters in seizing the moment.

Instruments mostly agreed to in North-South negotiations. I am going to use here the United Nations Conference on Trade and Development (UNCTAD) Code of Transfer and Technology as an illustrative example of several such initiatives which have gone through prolonged negotiations up to now. These negotiations involved, as a minimum, three distinct phases:

- formulation of the proposals by a Group of Experts convened by the Pugwash Conference in 1973;
- their acceptance – a very long drawn-out laborious process involving many modifications – by the developing countries themselves; and the endorsement of the modified proposals at the highest political levels of the Ministerial Meetings of the Group of 77 and the Summit Conferences of the Non-Aligned countries; and
- the negotiations, again very long drawn-out, with the countries of the North, which have brought the proposals to near universal agreement except on a few points. The efforts involved were simply enormous.

I am a bit disturbed by those who are busy writing epitaphs of the New International Economic Order (NIEO). Of course, the South has not converted the North to accepting it in full. But it should not be forgotten that the United Nations Declaration on the NIEO was not altogether unlike the "Declaration of the Rights of Man and of the Citizen", which formed the Preamble of the

New Constitution proclaimed by the French Assembly on 14 September 1791. Both called for a New Order: the French Declaration for restructuring relations between the citizen and the state; and the UN Declaration for restructuring the relations among the unequal nation states themselves. Both are part of the human heritage, impervious to death or dishonour.

With an unprecedented outburst of creative energy and sustained perseverance, the South has indeed brought together some two-thirds of the building blocks needed for the construction of the NIEO. Much has changed in the relations among nation States. The South can be proud of this achievement. Such an historic contribution can never be consigned to the waste-paper basket of history. It is just as immortal as the French Declaration on the Rights of Man.

Some countries of the North, particularly the most powerful ones, through their short-sightedness have during the 1980s developed cold feet over their own contributions to the construction of the NIEO. But that is no reason why the South's historic initiatives will ever go to waste. Let me illustrate this by the example of the UNCTAD Code on Transfer and Technology. All the articles of the Code have been agreed to, except for two: the prescriptive provision concerning restrictive practices; and the provisions on the applicable law for dispute settlement. By 1978, it had become obvious that these two articles were not going to be resolved easily. The only sensible course, it had become clear, was to work out a trade-off between the two. The trade-off would have involved a compromise by the Group of 77 on applicable law for a compromise by group "B" on restrictive practices. But these compromises could not be made in time. The negotiations have been stalled since then by the conservative governments which came to power in some of the countries of the North. And, unfortunately, the South by its equally short-sighted uncompromising attitude allowed this to happen.

I believe that the UNCTAD Code has relevance not only for relations between the North and the South, but also, equally, if not more so, among the countries of the South itself. Any expansion of South-South trade, industrial, financial and technological cooperation, is critically dependent upon overcoming suspicions and resolving the conflict of interests between the more advanced and less advanced countries in the South. Why cannot the less advanced of the countries of the South demand, and the more advanced countries of the South offer, to practice all the relevant proposals of the Code in their mutual transactions, without waiting for the Western countries' agreement to the draft Code? Such a step should at one stroke create an entirely different technological and economic climate for South-South cooperation. It would dispel the fear of the weaker Third World countries. It would demonstrate the sense of responsibility and maturity of the stronger ones. It would electrify the atmosphere and create a new ethos for South-South cooperation. In this period of all-pervading despair, such a bold step would generate the much needed new spirit, the new confidence for shaping the South's destiny. The world is waiting for a positive shock.

I have used the UNCTAD Technology Code only as an illustrative example of the major steps that could be taken by the South. Several other initiatives which have ripened through prolonged negotiations could be also considered in this context. An important element in all this is that full agreement on the proposals has been reached among the countries of the South. Moreover, most of the articles of these proposals have even been agreed with the North. Only a small step, centering upon a few issues dividing the North and the South but not obstructing South-South cooperation, need be taken. Is it purely Utopian to think that the time is indeed ripe for unilateral action by the South on these initiatives? Such action could take place on the solid foundations of the blocks of the NIEO already built. Several countries of the North may then begin to have second thoughts. Some of them may even join in. For the others, the South can wait. Times will change. And the South can spare the time.

New instruments for the South. The point mentioned above could also be examined from the point of view of forging new instruments for South-South cooperation. The successful negotiations of GSTP immediately comes to mind. This point will also apply to those proposals which require significant adjustments to serve as useful instruments for South-South cooperation. The time is ripe to begin such adjustments. This would then open up for acceptance many of the proposals which have formed part of the discussions on the establishment of the NIEO. The essential merit of this approach is not vitiated by the fact that all the governments of the North are so far not prepared to endorse in full these proposals. World reality is, however, undergoing major changes, undermining the frozen geographical blocs of countries. I am even inclined to think that once the South takes up unilateral initiatives, it would then mark a long step in the eventual acceptance even by the North, or by at least several countries within it, of these initiatives.

3.2 Discarding Geographical Proximity as the Corner-Stone of a South-South Cooperation

Up to now, most attempts in South-South cooperation have been based on geographical proximity of the countries concerned. This has a certain logic. Neighbours know more about each other's problems, and may, therefore, be more ready to find common neighbourly solutions for them. This line of reasoning had also been in part the basis of the establishment of the Common Market in Europe. But this approach has two difficulties: neighbours not only know each other, but have usually been in conflict because of matters such as national boundaries and minority ethnic groups; and these conflicts have often been aggravated by widely different levels of development among the neighbouring countries. In consequence, the less developed neighbour has been sometimes not only jealous but actually very suspicious of the advantages that the more advanced neighbour might gain at his expense through such cooperation. These two difficulties have been jointly responsible for destroying, or fundamentally weakening, the attempts that have been made in the past to promote cooperation among geographically contiguous groups of countries. Perhaps a wholly different approach might be considered now. That would involve jumping over the barriers of geographical proximity. The sharp decline in the cost of transport of goods, and the increasing ease of transmission of technology and informatics, would facilitate such a jump.

Looked at this way, cooperation does not have to be limited to neighbours alone, or even countries in the same continent. It could actually jump over the land and water barriers and embrace countries which are, to begin with, at somewhat similar levels of development, and are engaged in tackling similar problems. They would, therefore, be facing similar difficulties. They will not be thwarted in their efforts by the usual conflict of interests between the more and the less developed countries in the South, which has wrecked so many of the past initiatives. Cooperation along these lines has been earlier attempted between Egypt, India and Mexico. There is a need for distilling the lessons of this experience, which can serve as a guide for the future.

3.3 Diversifying Destinations of Cooperation

Since the end of the Second World War and the achievement of independence from colonizers, some progress has been made by developing countries to change the pattern of partners in their economic relationships. The progress, however, has remained rather limited. The origin of imports and the destination of exports still continue to be more or less limited to the earlier pattern. Despite some progress in South-South cooperation, the major partners for trade, finance and technology exchanges of the South, even today, are the Western countries.

There are three important developments which, if properly exploited, can help bring about a major reorientation of the past pattern. First, within the group of developed countries, there has been a big change in the balance of economic power. Western Europe and Japan have gained significantly, and the United States has correspondingly lost its pre-eminence. Competition among them has thereby considerably increased. This has opened up possibilities for exploiting this competition, bordering on fierce economic rivalries. In order to benefit from this new opportunity, the information base and the negotiating capabilities of the developing countries would require to be significantly strengthened.

The second factor relates to the opening up of the socialist countries to the world outside. These economics have raised their domestic output and demand significantly over the last 40 years. But the demands have been met mainly from domestic sources. There is a drive for greater efficiency reflected in an increasing desire to obtain the needed supplies from competitive markets. Here, the developing countries have indeed an immense opportunity for the expansion of their exports, not only of traditional primary commodities, but also of manufactured consumer and capital goods. But this requires working out efficient institutional and other arrangements which can supply the needed information, and overcome the barriers of language and custom.

Thirdly, similar changes have also taken place among the developing countries themselves. Some of them have advanced faster than others. These differences have often been deplored. But what has happened conforms to the past pattern of uneven development among countries. That is in a sense inevitable. Such uneven development, however, also raises the degree of complementarity between these countries. It thereby opens up new possibilities for enlarged economic exchanges. Here again, as in the two developments mentioned above, a whole series of new attitudes, approaches and institutional structures have to be designed to draw maximum advantages out of the uneven growth. Particularly important in this connection is ensuring that the more developed countries do not try to gain unfairly at the cost of the less developed countries in the South.

The emergence of these vast changes in the competitive economic strength of nations could contribute towards positive progress only if the developing countries were to have at their disposal appropriate institutional structures to benefit from them. But most countries in the South are small and weak. Their skill and resource levels are low. They would, therefore, need new forms of institutional arrangements to assist them.

3.4 Enlarged Technlogical Cooperation

It is now almost universally recognized that technology is the master key to unlock the closed doors of the development process. But the recognition is much more recent in origin than that applying to trade and financial exchanges. A very extensive network of private and public institutions has grown up over the centuries to foster trade and financial exchanges among countries. There has also been a parallel development of norms and standards and established custom for promoting exchange of goods and finance. But there are no similar institutions and norms for technology, the relative newcomer on the world stage.

In the early concerns about technological exchanges, great attention was paid to restrictions on access to proprietary technology – patents and trade marks and know-how. These were then the most visible obstacles in commercial exchanges. Moreover, the level of skills in the developing countries was generally too low to exploit, or to build upon, publicly-known technological innovations. It would seem that two new factors have been significantly responsible for altering this past framework. First, the consciousness about the importance of acquiring the right technology at the right prices, and under the right terms and conditions, has now spread widely. The developing countries now know much more about the channels through which technologies are transferred, the relative strengths and weaknesses of these channels, the crucial importance of careful negotiations, the need for proper assessment of probable losses and gains in such exchanges, the importance of adjusting laws and regulations on patent and trade mark systems to promote national development and not just to grant monopolies to the outsiders, and the establishment of national and international norms and institutional machinery for such exchanges. The countries of the South are no longer like foreign tourists buying souvenirs abroad. The sustained efforts of the Group of 77, often channelled through UNCTAD, have been of critical significance in bringing about this change. While the first factor relates to the change in the general climate, the second factor is of immediate operational significance. That concerns the explosive expansion in the South of skill formation in general, and education in particular. In consequence of its educational sprint since 1950, the Third World has equalled *on a per capita* basis the comparable levels of primary school enrolment, reached the 1950 levels of secondary school enrolment, and nearly attained the 1950 enrolment in higher education in the developed countries, both in the West and in the East.

Particularly impressive was the rise in enrolment at the tertiary level in universities and institutes of higher learning. It grew at 10 per cent per year, rising over 25-fold in terms of numbers. Even when the quality of training left much to be desired, the sheer size of the number is formidable. Out of some 50 million students in higher education in the world, over 20 million are now enrolled in the Third World. The South now accounts for 4 out of every 10 students in higher learning. In 1950, this ratio was only around 1 to 10. Whatever some people may prefer to say about the widening technology gap between the developed and developing countries, there is one area in which the gap has narrowed – that is, in relation to the sheer size of numbers in higher education.

This expansion in skill formation in the South has now created the bases for utilizing much more effectively all the technological knowledge that exists as a public good – that is, not subject to private proprietary control. The universities, technical institutions, the research and the development laboratories, the journals and books in libraries, the international conferences on special subjects, the widening contacts among specialists, the rapid ease with which information can be gathered, stored and exchanged for immediate use – all these phenomena now make it possible to draw the maximum benefits from the ever-enlarging storehouse of world technology. To these must be added the technological innovations by the national, and multinational R & D bodies. Such technologies have generally been turned over to the public domain. The trained men and women who can open the doors of this storehouse, undertake the required engineering, and design and manufacture the needed machinery and equipment, are all now there to accelerate the technological transformation of the South. An analysis of those developing countries which have progressed more rapidly than others, shows beyond any doubt that the formation of skilled manpower or social technology, was the most important factor in their rapid transformation.

This somewhat long discussion is meant to underline that the scope for enlarged South-South technological cooperation is now simply immense. This is the most significant untapped resource at the disposal of the South. Such South-South technological cooperation could embrace all aspects of technology – cooperation in training, in research and development, in design and consultancy, in plant operation and marketing, and in joint research on classical and new technologies.

It is particularly important to remember that the new technologies are relatively more easy to cope with, imitate and reproduce. They make fewer demands on material inputs and capital. On the other end, they are highly skill-intensive.

It may be mentioned in passing that the ease of absorption of these frontier technologies is the most strategic reason why the developed countries have started their offensive, in the Uruguay round of negotiations in GATT, to strengthen their monopolistic control on all forms of intellectual property. But, then, that is precisely the reason why the Third World countries, as has been emphasized by the South Commission in its statement on the subject, should not allow this new window of golden opportunities to be closed.

3.5 World Bank for Technological Cooperation

The idea of a World Bank for Technology has been around since UNCTAD III in 1972. But it has never been fully explored. Perhaps the conditions were not ripe in the past. The scope for using the stock of technologies in the public domain was not sufficiently known. Nor did technology occupy the centre of the stage as it does now. Nor had the Third World developed the critical mass of skills needed to exploit this stock. Nor could the proposal muster the support of the developed countries. As discussed above, several of these obstacles are less strong now. President Gorbachev, in his address to the Chinese people in Beijing in May 1989, offered his support for such a scheme for sophisticated technologies.

The functions which such a Bank could perform require careful discussion. Among them, a prominent place should be assigned to the following:

- bringing together and transferring at cost the technologies developed mainly by multilateral bodies, such as the World Health Organization, the Food and Agriculture Organization and others; by international and regional research institutes of which a sizeable number have now begun to function; by publicly-financed research and development bodies at national and multinational level;
- transferring at concessional rates other much needed technologies which could be obtained for use in the South; these operations could follow the techniques with which we are now already familiar through the General Scheme of Preferences in Trade, and the soft loan facilities of the World Bank and other financing institutions;
- promoting joint research and development designed to solve the critical problems faced by the South;
- inviting cooperation of scientists and technologists on a global basis to assist in the creation of a world pool of technologies.

Such an institution could be a landmark in international technological cooperation. It could foster R & D. It could act as a source for technologies in the public domain. It could assist in transfer at concessional rates, thus strengthening the process of progressive decommercialization of technology exchange.

It could enlist the cooperation of scientists and technologists to add their own contributions to the common heritage of mankind.

3.6 The Public Sector as an Instrument of South-South Cooperation

An important instrument at the disposal of the Third World to foster South-South Cooperation is their ready-made public sector. For several historical, economic and operational reasons, all the countries in the South now have a formidable public sector. It consists of enterprises at the centre and at state levels, engaged in building up the infrastructure, providing essential services, manufacturing critical goods, distributing and selling and buying in domestic and external markets. The demands of the public sector are indeed very large. But they have not been so far sufficiently exploited for fostering South-South cooperation.

Recently, much is being heard about the inefficiency of the public sector. Certain vested interests have called for dismantling the public sector and the para-statal enterprises. They want instead privatization. This is not the place to discuss this issue in any detail, except to point out that the conditions in the South are different. They require a strong and efficient public sector.

The very large public sector demand is a very potent instrument for furthering South-South cooperation. An efficient harnessing of this demand could at the same time serve as an instrument for overcoming some of the inefficiencies and weaknesses of these enterprises. Here is an area in which the instrumentalities which have already been constructed are not being used purposefully to promote South-South cooperation. The possibilities of their more effective use need, therefore, to be examined in detail, drawing upon the past contributions of several bodies such as UNCTAD.

4. Appropriate Institutional Structures

We have examined several areas in which greater South-South cooperation could be promoted. Throughout the analysis, we have stressed that appropriate institutional structures are needed to derive the maximum benefits from these new openings. General principles are fine. But they need institutions to assist the practitioners.

Several institutions dealing with sectoral as well as overall concerns now already exist in the South. The country coverage of these structures is often limited. But they are there. There is, however, an absence of an overall apex body which could in a very rough and ready manner offer useful advice and help to the countries of the South. Suggestions have been made for some institutions which should begin to perform the functions similar to those now carried out by the Organization for Economic Cooperation and Development (OECD). It should be recognized that the needs of the South are not exactly those of the North. This would have to be taken into consideration in the future establishment of any such structure. The new body will have to address itself not only to economic but also technological and social issues. For that reason, whatever name is chosen, the new entity should have wide enough scope to deal with all relevant issues. As an initial suggestion, the name could be something like Organization for Economic, Social and Technological Cooperation and Development of the South. But it is long and clumsy. Something shorter and easier to remember will have to be found.

The idea of establishing such a structure has been around for a long time. But there has not been much progress towards realizing it. There are several rational as well as irrational fears about its role. These issues need to be sorted out. Any proposal for such an organization must start on a relatively modest basis. But it should not be so modest that it would prove wholly ineffective. It should take into account all the building blocks that exist at the inter-regional and sub-regional levels. It should begin to serve as a bridge between them. It should grow slowly, depending for its expansion upon its demonstrated ability to assist the South.

There is one factor, however, that would outweigh most of the hesitations, or the irrational fears. It is the asymmetry among the South countries in terms of the numbers of their nation states and their populations. Table 1 brings out this symmetry.

Table 1. Population and Country Composition of the South

Population	Number of Countries	Total Population (millions)	Share in Percentages
Over 100 million	4	2000	60
40 to 100 million	11	650	19
10 to 40 million	22	400	12
Under 10 million	91	300	9
Total	128	3350	100

As can be seen from the Table, the South consists of a few countries with very large populations, and a very large number of countries with very small populations. For instance, only four countries, each with over 100 million people, account for 2000 million or 60 per cent of the population of the South. At the other extreme, there are 91 countries, each with a population under 10 million, which together account for less than 10 per cent of the South population. In fact, India or China alone is larger in population than 113 other countries together. It is this asymmetry between the number of states and size of their populations which constitutes the compelling reason for the establishment of an organization of the South. In real life, the small and the weak need much more help from such a body. The strong will prove short-sighted if they do not respond to this need.

5. Conclusion

I conclude by stating that these proposals may appear to some as Utopian, as dreams for years not yet come. But our dreams of today have so often been realities of tomorrow. The worlds of the West, the East and the South are in desperate need of dreams. The task of scientists and politicians is to build bridges between the dreams and reality. The current mood of despair would, I believe, pass. The dawn of tomorrows is already breaking out on the distant horizons. That is the conviction, or belief as some may call it, which has guided me in outlining these approaches. I hope that at least some of them may be useful.

A Decade of Intended Cooperation in Science and Technology – The 1979 Vienna Programme of Action

Essam Galal

1. Introduction

The *Vienna Programme of Action* was approved by member states of the United Nations in 1979. It prescribed a comprehensive programme for the application of Science and Technology (S & T) to development, particularly of developing countries. It involved actions by developing countries, as well as cooperation with developed countries and the UN organization. It created a machinery of a UN Intergovernmental Committee, an Advisory Committee, and a Centre for S & T for development, but a fund to finance the implementation of the programme did not materialize.

After a decade, the UN reviewed the progress achieved. All countries, regions and UN agencies were asked to participate in this review. A series of regional meetings culminated in an interregional meeting, sponsored by the UN Centre of S & T for Development, and the German Foundation of International Development. The meeting, held in Feldberg, FRG, was attended by experts from members states, regional organizations, UN agencies and members of the UN Advisory Committee on Science and Technology for Development. As the outgoing chairman of the Committee I gave the keynote address, in which I assessed the past experience. This paper contains the substance of my address.

2. Historical Background

Our concern with the role of science and technology for development in the late 1960s and early 1970s, reflected our experience in developing countries with newly-launched industrialization programmes in a period of positive growth. No doubt, in retrospect, an element of naivety can be discerned in our visions, but even with hindsight, we can claim we were not fools.

We must first indicate the components of the package that reflected this vision, of which the Vienna Programme was only one and not the earliest component. And let me stress that it would be misleading to attempt an assessment based on only one component of the package, without the politico-economic realities that instigated it. The most prominent points were as follows:

- Need of political unity and clout of the Third World. I believe it is accurate to claim that without a perceived political clout, it would have been inconceivable for us to call for mutual dialogue with the North to restructure the many foundations of international economic, science and technology relations.
- The New World Economic Order reflected our realization of the necessity to rectify the marked imbalance in accessibility to trade, investment, resources and opportunities, as a prerequisite for a viable development.
- The Code of Conduct on International Transfer of Technology reflected our early realization that while intergovernmental cooperation may be effective in answering some of our most urgent scientific requirements, in a free market economy it is the private sector which holds the taps to most vital requirements in the field of technology capabilities.
- The preaching by UNESCO of supply-oriented strategies of R & D promotion, which is now often criticized, in reality reflected objectively the demands of the period, when most developing countries needed to build their national infrastructures from scratch, not only in science and technology but also in the wider state-creating endeavours.
- The UN system and agency harmonization project which preceded the Vienna programme, reflected our realization that the fragmented non-coordinated and not always S & T-based multitude of UN inputs in developing countries, have a potential for an organized and systematic contribution that can multiply the returns of its meagre resources.
- The development decades, with their emphasis on the changing needs, opportunities and environment of development, again reflected our holistic appreciation of the dimension of our problems, and the requirement of global prosperity and peace.

With this historical perspective in mind, I consider that the facile pronouncement of a decade of failure of one particular programme fails to take into consideration the historical dimension of the masrive transformation involved and the multiplicity of its facets.

In reality, the shortcomings in the implementation of the Vienna Programme of Action reflects the regrettable, but understandable, current difficulties in realizing the vision of the whole package, and how necessary the goals are, and will continue to be.

It is counterproductive to try to allocate responsibilities for this unfortunate situation. All partners share the responsibility, but I want to emphasize our own responsibilities.

We have not served our cause by mixing our priorities, mismanaging our development, neglecting and squandering our resources, missing our opportunities and failing to formulate, sustain and implement our policies. In fact, mixing priorities and mismanagement started at Vienna, when we supported substituting the instruments (the Fund and institutional arrangements) for the goals of sustained mobilization policies and linkages. Of course, financial flows had higher visibility and comfortable accountability – however marginal their role was and still is – than flows of information, capabilities and opportunities. Unfortunately these are the real decisive flows.

It is also true that not all developing countries lived up to the implied responsibilities of Vienna. These implied responsibilities included assigning a high and integrated priority in their development strategy, if and when acquired, building up the human institutional resources and allocating the necessary funds and responsibilities.

I claim, however, that the *coup de grace* to the vision of Vienna was delivered not to science and technology but to development. After all, international cooperation in Science and Technology for Development can at best be only a seed investment. Seeds do not grow in parched earth. The earth of development was severely parched during the last decade.

3. Reversed Development After Vienna

The media in which science and technology were to flourish and contribute to development, suffered a tragic reversal in the 1980s. For indebted countries, the average per capita income decreased by 1.3 per cent annually, over the period 1980–89, while projected growth estimates (1987–1995) vary from +1 to –2.5 per cent. For Africa the reversal was even harsher. Actual per capita growth (1980–87) was –2.7 per cent, even though population growth reached 3.99 per cent. The African debt in 1986 reached 263.3 per cent of income, and 76.2 per cent of the GNP, compared to 55.9 per cent in the highly indebted developing countries.

The projected growth of national product in Africa during 1986–1990 is –4.6 per cent, while the annual growth rate of debt has reached +6.8 per cent. A more illustrative exposé is given in the UNCTAD report. In developing countries, the decline in terms of trade was – 20.0 per cent in 1986, continuing the trend of preceding years. Weakened prices resulting in estimated loss of $80 billion despite an actual output growth of +3.2 per cent in developing countries.

To sum up, the purchasing power of export during 1980–1987 was:

Industrialized countries	+45.0 per cent
Africa	–50.0 per cent
Latin America	–26.0 per cent

The World Bank emphasizes the process of "adjustment" with the developing countries as a pre-requisite for successful management of their economies. This process implies, no doubt, sufferings in the developing countries, sometimes amounting to unanticipated shocks. It implies that for developing countries successful management requires "adjustment", presumably meaning accepting and paying the price of transmitted shocks. It shows that the assessment of the Vienna Programme achievements cannot honestly be carried out in the context of its proclaimed aims, namely Science and Technology for Development. Realistically, it has to be carried out in the context of science and technology for reverse development. What is more disheartening is that this tragic situation is predicted to apply at least for the next half decade.

4. Flow of Resources

In the light of the $80 billion loss of export prices and the $100 billion cost of debt service in 1986, it is a euphemism to discuss the role of the lost Fund of $0.5 billion, or even the hoped for $2–3 billions of aid from all divergent sources. What we should be evaluating is the contribution of the reverse flow, from developing countries to the advanced economies, and the remarkable progress in the application of science and technology to their own development. Probably a more disturbing reality is that under these adverse conditions the trickling "aid" from advanced countries is playing a critical role and is likely to continue to do so. Under the reverse-development situation a profitable employment is not likely to be feasible.

The growing diversity and multiplication of its instruments without harmonization could still be more confusing. Its orientation is likely to continue on the current rack, i.e., crisis management (sometimes misnamed basic needs), while sustainable growth and capability acquisition may continue to play second fiddle to the necessities of survival, if at all. The growing protectionism and market competition in the North may further increase the politicization and commercialization of cooperative resources, further compounding the difficulties of development policies under conditions of crisis in developing countries. Science and technology resources flowing from the private sector do not seem to be free from similar difficulties. The US Department of Commerce estimated that US multinationals spend 90 per cent of R & D expenditure at home, 9 per cent in other industrial countries, and only 1 per cent in developing countries. There is evidence that this 1 per cent is mainly concerned with quality control of local adjustment and very little of it goes towards building national capabilities.

Taking into consideration the situation of official aid and the growing importance of the private sector contribution in implementable technology, and the freezing of the negotiation on the Code of Conduct of Transfer of Technology, this is likely to represent a growing problem for developing countries and for the realization of the goals of Vienna.

However, developing countries very "effectively" contribute to their own difficulties. Squandering of resources and missing of opportunities is a popular hobby for some of them. Take the example of the Arab World. Acquiring a strong negotiating and most favoured position should have come naturally to many Arab countries, considering that their share of the total exports of major technology exporters in the West was substantial. It ranged from 19 per cent to 56 per cent for different technology suppliers in 1982, and totalled $40 billion. Yet, technology assimilation does not seem to have improved. The same technologies in the petrochemical industry seem to have been purchased several scores of times. Egypt with a technological and industrial experience of over a century and a half, had finished the first Egyptian design, manufacture, construction, and operation only in 1987.

The flow of resources through trade, as previously indicated, did not bode well for development in the last decade. There is no indication that the share of developing countries in world trade is going to increase in the near future, neither is the volume (because of increased protectionism and market blocks), nor in regard to prices, because of the prevalence of alternative materials, energy, resources and saving, as well as less labour intensive technologies.

The flow of resources and effective support from the UN system, the third partner of the Vienna Triad, has to be viewed cautiously, notwithstanding the recent revival of some of the UN security functions. In the field of development nothing has emerged yet to indicate the imminence of a more effective role for the UN. In fact, the growing emphasis is on bilateral channels. This is coupled by growing politicization of presumably non-political multilateral agencies, like the World Bank and the Monetary Fund, and the depoliticization of basically politically-influenced agencies, like UNCTAD and UNESCO. These trends do not augur well for developing countries' expectations from the UN system as expressed in the Vienna Programme of Action.

The role of human resources in the evolution of utilization of science and technology for development seems to be exposed to a crisis in quantity and quality. The brain drain, a subject of much denunciation in Vienna, has been replaced by policies of active exports of unemployable brains by developing countries. Frequently, this does not reflect any real excess of human resources as compared to development needs. It simply reflects the non-growth of the employment market as a result of diminished investment. The qualitative crisis is even more menacing where the economic crisis is denying vital funds on education at all levels, to R & D and to further education. The dismantling of the rudimentary capabilities created over years is not a remote possibility in the poorer nations. The retardation of national progress in science and technology capabilities, even in the more favoured countries, should arouse grave concern.

5. Other Added Complications

The necessity to "readjust" frequently development strategies in response to transmitted "shocks" from the global economy, due to security emergencies of natural disasters, have often mixed the priorities and disrupted the programmes, resulting into substantial losses. The overriding demands of "survival" are often pushing the need for development to a back seat.

The current technology explosion, while in the long run may offer windows of opportunities for some developing countries, has presently eroded the marginal advantages that some developing countries had with new science and capital intensive technologies. Cheap labour, or available resources, such as copper, lost much of their insurance value.

The erosion of the presumed interdependence that fuelled many of the reformist ideas of the sixties – as the technology-based power restructuring crystallized – has ushered in the era of neo-dependence. Its manifest economic and military dominance is likely to be propagated for decades through science and technology dominance.

The 1990s may also see the eradication, or at least the reduction, of the "exceptional" opportunities for new extension economies, as in Singapore and Hong Kong, or in the strategically-favoured Taiwan, as the economic and military need for new candidates is negated by new technologies, economic competition and detente.

This attempt at an objective assessment of the experience of a phase of global change should not be mistaken as an expression of pessimism or defeat. The goals of the vision of the 1970s should be seen within the context of a realistic historical perspective. While change may fall short of aspiration, the significance of the achievements, however modest, should not be underrated. If it takes a couple or more of decades to bring us nearer to the fuller realization of the vision, that would be no small achievement. After all, it will be truncating the three centuries that the advanced economies needed into modest decades. For one thing, the recognition of international partnership, however reluctantly, and the maintenance of international dialogue, however fruitlessly, are no minor achievements. The creation of instruments of change, however impotent they may currently be, is a significant gain.

On the side of the developing countries there has emerged some very promising progress, too. Heightened awareness, diverse capabilities, mounting experience, growing institutional structures, programmes and missions, mastering of organizational, planning and management abilities and experience. Access to information through wider means of communication, has raised expectations and repeatedly contributed to the decline of militarism, the growth of participation, the rejection of corruption and mismanagement, and the safeguarding of national resources and rights.

On the developmental front there are also some promising developments. There is more readiness for objective and realistic sorting out of mixed priorities. There is more readiness for a flexible approach to the choice of appropriate and adjusted economic and institutional models and explorational mixes of these models that suit particular situations and conditions. Very significantly, we have seen the growing realization of the need to differentiate between the realities of current scientific dependence (as illustrated by the current expenditure of $400–500 billion on R & D in the advanced countries), and the unjustifiable technology estrangement. Rewards have been achieved by the efficient execution of technology management policies. More and more developing countries are graduating from the supply-oriented science and technology policies, that were justifiable in their infrastructure building stages, to a dynamically adjusted mix of supply and demand promotion and response policies. These policies are furnishing the bases for more effective technology transfer, international cooperation, and gradually initiating technology generation.

6. Future Orientation and Choices

The road is no doubt long and hazardous. Yet the openings are there for those developing countries which will succeed in developing the management capabilities of the most dominant feature of the coming decades, namely change.

Some of the choices towards that end are:

- In a technology-based global power restructuring, science and technology cannot any longer be treated as the paraphernalia of development. Science and technology have to be treated as independent national priorities alongside the basic needs of health and education.
- With the accelerated tempo of scientific and technological change, staged strategies are no longer justifiable. A multidimensional strategy for science and technology growth must simultaneously target infrastructure building, capability acquisition, technology management and generation.
- *There is a need for*:
  more flexibility of planning and adjustment of development and economic models, market practices and innovation, to respond to global market trends and dictates;
  more clarification and rationalization of development goals, and more mobilization of assets and research, including harmonized international cooperation in the service of these defined goals;
  higher recognition of the importance of South-South cooperation, coordination and market consolidation to face mounting protectionism and market blocks.
- In spite of all disappointments and difficulties, channels and instruments of international dialogue and cooperation must be safeguarded, not only in the service of current needs but also in the service of the still more valid long-verm goals of the vision.
- We must ensure mobilization and commitment across the board, participation, opportunity, information, incentive and management to narrow the efficiency gap.

In the second decade after Vienna, and despite all the current and impending difficulties, there are very promising opportunities. Opportunities for those who can, first and foremost, at the national level, lay the appropriate foundations.

Only those who can will get more capabilities. Only those who know will get more information. Only those who plan and manage well will get better returns. Only those who venture efficiently will gain bigger markets and generate more capital. The only strategy for the coming decade is the strategy for acquiring these self-propagating abilities, the most basic of the basic needs.

The Roles of Transnational Public and Private Mechanisms

Hans-Peter Dürr

1. Introduction

Our means of destruction – military and non-military – have risen to a level to threaten humankind as a whole. Our means of transportation and communication have shrunk the globe to a village. But our mentality, our ways of approaching problems have not changed accordingly. We are still trying to solve problems with ideas and tools of the past. We are still thinking in terms of nationalities, sovereignties, borders, separation from others, without taking into account the fact that the most pressing problems of our time are global in character which ignore all these distinctions. We have to change our way of thinking if we are to attack these problems properly and successfully. We need new mechanisms which shake off disciplinary limitations and overcome national borders. These mechanisms must be installed by governments and other public institutions and organizations. But most likely they have to be initiated by private actions, committed individuals, project-oriented groups and non-governmental organizations. All forces have to be mobilized to meet the global challenges.

In this paper I will first deal with the problem areas which most urgently require East-West as well as North-South cooperation. The character of these problems and their underlying causes will be indicated. For their successful solution certain preconditions must be fulfilled and a sequence of measures have to be taken. Secondly, I will consider whether public or private mechanisms, or a certain combination of both, will be most appropriate to meet these needs. Finally, I will give some concrete examples of transnational private initiatives presently concerned with solving global problems.

2. Global Challenges

All global problems derive from local causes, but not all local incidents develop global features. For this to happen certain additional factors are necessary, in particular, a capability to spread and a capability to become amplified.

The arms race with the potential of an all-out nuclear war still represents the greatest danger to human civilization and the human species. To find ways to break this vicious circle and to settle future conflicts by non-military means is one of the greatest challenges of our time. Nevertheless, we do recognize today that there are other global problems of equal urgency and gravity threatening humanity: the degradation of our environment by the rapid industrialization; the accelerating consumption of non-renewable resources, in particular of fossil fuels; and the increasing economic imbalance between North and South, with the concomitant impoverishment of the majority of a growing world population. Pugwash, although still mainly concentrating on the important military security issues, has turned to these other relevant problem areas in the past. In fact, during its 38th Conference in Dagomys, with the overwhelming consent of the participants, the Pugwash Council adopted the Dagomys Declaration "Ensuring the Survival of Civilization" in which – in the spirit of the original Russell-Einstein Manifesto concerned with the risks of nuclear war – scientists were called upon "to expand their concerns" to a broader set of interrelated dangers: destruction of the environment on a global scale and denial of basic needs for a growing majority of humankind [1]. It is my conviction that concentrating on this broader set of global dangers will also greatly help to resolve the military issues, because the necessity of global cooperation in these cases is even more obvious. It is the environmental problems which should lead our way. Because of their comprehensive character, their solutions will not only reconcile man with nature but also man with his fellow-man.

Our biosphere and ecosphere represent an extremely complicated and ingeniously interlinked system of regulatory circuits which are designed in such a way as to absorb practically all external impacts on the system and to keep it in its state of dynamic equilibrium. The robustness of the ecosphere is very impressive and can only be understood as a product of its billion-year-long evolution under changing conditions and corresponding continuous adaptation. This relative robustness of our ecosystem gave us the impression that our environment is practically invulnerable: an infinite reservoir of resources which can be quarried without limitations, and into which we can dump everything we do not need without any repercussions. But today we realize that this is not true: our ecosphere is vulnerable and finite. In particular, by tapping the immense amount of solar energy trapped over millions of years in the form of fossil fuels, and the highly concentrated energy of heavy atomic nuclei baked in the very hot early stages of our universe, our technical means of amplifying our forces have grown to such an extent as to compete with the natural forces of our earth. As a consequence, our impact on nature can no longer be considered as negligibly small. Therefore, we may very well imagine that human influence may exceed the critical value and hence may push the ecosphere out of its present state of equilibrium, causing a "phase transition" to a new state, in which the general conditions can be so dramatically different as to become incompatible with the existence of the human species and other higher forms of life.

The dangers of forcing our ecosystem into a different state of equilibrium arise not only from strong actions but, more subtly, may also be caused by new agents for which our earth was not prepared on its evolutionary path, such as heavy metals, which we dig out from deep, hitherto sealed off parts of the earth's crust, or chemicals synthesized in our laboratories. These new agents act like new virulent viruses from which our ecosystem is still not immune. Global problems, therefore, may arise from our activities, either by exceeding the "elasticity limits" of the ecosystem in exerting brute force without constraint, or by ignoring the sensitive points of the ecosphere.

The process of amplification is strongly pushed along by our tendency to seek efficiency, that is, by our tendency to produce a maximum output of a well-prescribed type by a minimum of input, and organizationally supported by centralized structures. In contrast to nature, which by the peculiarity of the evolutionary process tries to maximize the number of options in order to optimize flexibility and hence long-term survival, human technology emphasizes high-performing one-track solutions with speedy short-term returns. To achieve maximum amplification strong feed-back mechanisms are employed. They favour the system of developing an *eigendynamics* which, in severe cases, denies any external steering and, in general, takes a destabilizing course with potentially catastrophic consequences. Some of the global problems facing us today may be already "out of control" in this sense. Therefore, a remedy in these cases will be extremely difficult to find. But before we can hope to steer such a system in the right direction we have first of all to find a way to bring the system back into a controllable state.

3. General Goals

An ecosphere which is finite and essentially a closed system requires very different behaviour on our part than when one is dealing with an infinite, open system. In an infinite, open system, a life-style concentrating on prospecting and exploiting of resources, and expelling debris and wastes at an increasing rate, appears quite appropriate and most convenient, but in a finite, closed system this life-style cannot be sustained in principle, and in the long run will be lethal.

The finite, closed system requires that not only a segment of the production cycle, from production to disposal, profitable for us is considered, but that the full cycle, including the missing part, with wastes being transformed into new sources of production, has to be taken into account. From the Second Law of Thermodynamics we know that such cycles cannot be sustained for ever without external support, because syntropy, the negative of entropy, will necessarily decrease with every transformation. In fact, the faster we go, the more we speed up the transformation process. This, however, does not spell disaster, because fortunately our ecosystem is not completely closed; it receives a daily allowance of syntrophy through the incident of sunrays. It is this syntropy which supports all life on earth and it is this "capital' we can tap, without creating debts, to drive whatever transformation cycle we like for our benefit. Instead of a supply-oriented economy based on human "demands" – which in the industrialized world have very little to do with actual human needs – we have to aim at an ecosystem-sustaining economy in which – as in every sound economy – consumption is never allowed to exceed normal income.

4. Obstacles to Transformation

The goal of establishing an ecosystem-sustaining economy is rather obvious. But the obstacles to reaching this goal appear prohibitive. In contrast to military security problems, where at least in principle a solution – namely general disarmament – is apparent, no such simple solution suggests itself in case of the ecological problems – except if one aims, in direct analogy to the military issues, at an extensive reduction of production and consumption in the industrialized countries. Such a proposal seems at present unacceptable to most people in these countries. Hence one seeks to attack the problems in the usual superficial fashion: one tries to cure the symptoms instead of eradicating the disease, that is to repair the damage rather than to avoid it. In economical terms this procedure seems also to be more profitable, at least on a short-term basis: costs of repairs occur later, whereas avoidance – although in general much cheaper – requires investment now. In addition, causing damage and repairing it afterwards demands double effort and hence offers the opportunity for double profit both entering additively in the GNP which is still considered to be an indicator of our standard of living.

To remedy the ecological situation does not require a return to the Stone Age, as often proclaimed, but a return to reasonable behaviour, namely to produce all the goods and services, which we consider really essential and valuable for our lives in the most efficient way, that is with a minimum of syntropy. This, at the same time, will also turn out to be the most cost-effective way of production, a requirement which appears to coincide with our present approach to economic rules. But, in fact, it does not so coincide, because our present economic bookkeeping takes into account only a very minor part of the real costs. It neglects the social and ecological secondary costs, to be covered by the present or future taxpayer, the exploitation of the Third World, and, last but not least, the unaccounted loans from our generous creditor Earth which we simply regard as gifts. An honest calculation would clearly disclose that our productivity, which relates the intended benefits (serving our basic needs) to the total expenditure (including the used-up non-renewable resources) will be miserably low, contrary to our usual ways of accounting. We are probably not too far off the truth if we claim that the accumulation of any large wealth has little to do with high proficiency and accomplishment but simply reflects a skill at exploiting most effectively one's fellows or nature. The only gain of wealth which is strictly non-exploitive is directly or indirectly connected with the syntropy offered free of charge by the sun, and this gain is limited and highly dispersed. How can we create an economy which takes into account these basic realities? How can we eliminate the fundamental error in our present economic calculations, which assumes that interest gained by capital is real and not mainly caused by a clever shift of unaccounted assets of other people and of nature to the accounted ones of the capital owner?

To change this established pattern we have to break up the enormous and highly centralized power structures which use this system to their own advantage. East and West, although very differently structured, face a similar dilemma: both have to search for less centralized structures – politically in the one case, and economically in the other case – to establish a sustainable economy. That none of the developed, industrialized countries have as yet a valid answer for the future, may be considered unfortunate, but this precarious situation does offer, to my mind, a real chance for East and West – taking into account some of the rich experiences of sustainability by the South – to search jointly and cooperatively for a solution of this vital problem. The roles of public and private mechanisms have to be designed and measured with this goal in mind.

5. General Mechanisms for Change

Attacking global problems is very difficult:

- they are huge in dimension, exceeding by far our individual strength and imagination, with the consequence that they largely frustrate and even paralyse our own involvement;
- they appear intricately linked to everything and, therefore, offer no obvious leverage points;
- their most serious consequences may only show up in the distant future and their perception, therefore, requires deeper insight and broader knowledge;
- they are subject to a strong *eigendynamics* which does not respond to our usual static problem-solving approaches, if it can be influenced at all.

A necessary condition for any change is a learning process: a growing awareness of the various causes of the global problems and their inter-linkages. Learning processes, unfortunately, are usually slow as compared with the man-caused transformations of the ecosphere. Smaller and increasingly larger catastrophes, therefore, may have to become our most effective teachers, if we continue to behave as stupidly as we do now. We should make every effort to forestall them by our anticipating intelligence. Education will have to play an important role. The media – newspapers, magazines, radio and television – should help to transmit the relevant knowledge and to encourage people to engage themselves in the search for a sustainable future. They should make people fully aware of the dangers. But they should not leave them in a state of fear and agony. On the contrary, they should give hints at possible solutions and engagements. Clearly, intellectual insights into the problems will not alone suffice. We need appropriate tools which transform thoughts into concrete actions; we have to find the proper access to the problems and the right methods to attack and solve them.

To tackle successfully these problems we definitely need experts of various types to offer us the necessary insights and basic knowledge. But people at the grass-roots may be better guides to directions in which to go, because they are usually more aware of the interlinkages of processes and their holistic features. To weaken the destabilizing *eigendynamics* enforced by strong non-linear feedbacks arising from centralized structures, that is to get the system back into a manageable and controllable state, highly centralized but co-ordinated mechanisms have to be employed. This means that people themselves, in large numbers, have to take the initiative.

To attack global problems by means of globally-designed actions may be desirable, but I fear that in our political realities will prove rather ineffective. The disarmament discussions clearly reflect this point. The most obvious solution to the publicly-proclaimed goal of both sides to improve security, namely a deep cut in all armaments accompanied by a defence-dominated posture, cannot be successfully negotiated because of the selfish interests of the parties (do they really want disarmament or only pretend?) and their mutual fears. The "I-will-if-you-will" procedure is too slow and, in fact, fails if dynamics is taken into account. The only way it appears possible to make sufficient progress towards global change is to undertake unilateral steps – not necessarily in the sense of making concessions, but in the sense of adhering to a design by which severe disadvantages in the existing structure are avoided and a potential of great advantages in the long run is provided without suffering. This "I-will-act-independently-and-invite-you-to-join" procedure takes fully into account the positive effects of synergisms. To use another image: it provides a germ which can not only survive in a hostile environment but is able to grow and flourish and eventually will catalyse the necessary global change. It is the way that nature evolves.

Any change to become effective has to be shaped in a concrete fashion. The explicit formulation of very concrete in-between goals will offer the basis for constructive dialogues among people of different professional disciplines and different cultural backgrounds. Real solutions to our global problems will not be presented separately by experts of one sort or another, but only by an integrated effort of all of them including "thinkers" and "doers", intellectuals from the sciences and the humanities, writers, artists, ministers, managers, politicians and, last but not least, just people with common sense. In order to design, to initiate and to implement successfully these necessary changes, public and private mechanisms have to be activated. On the one hand, the novelty and complexity of the problems require imagination, ingenuity, flexibility, differentiating intelligence, pattern recognition and perseverance – all qualities which are best seen in private endeavour. On the other hand, the immensity of the global problems, the sheer size of the necessary efforts to make significant changes, requires tremendous financial resources, widely-extended and well-co-ordinated actions, and new general rules of conduct – requirements which can be hardly met without a strong involvement of administrative and governmental agencies.

In considering the advantages and shortcomings of public and private mechanisms, however, I have difficulty in making a meaningful separation between these mechanisms. It appears that a distinction between governmental and non-governmental mechanisms would be more appropriate, or between profit and non-profit organizations, or between centralized and decentralized institutions. I have finally decided to use the pair centralized-decentralized mechanisms rather than public-private mechanisms for the following discussion, because the different roles, as I see them, can be spelled out more clearly. But by going into details, and by illustrating my arguments with existing organizations and groups, I do realize that some arbitrariness is involved in attributing specific roles to any of these labels in one way or another. A private organization like, for example, *General Electric* and its transnational activities, has very little in common with a private organization like, for example, *Greenpeace*; they differ distinctly on the profit-non-profit scale but perhaps not so much on the centralized-decentralized scale. In their zeal to increase their own benefits, irrespective of the international "common good", a private multinational corporation resembles very much a national government although they rank as opposites on the private-public scale. Both are highly centralized, hierarchical and power-oriented. This may explain the reason for my particular choice.

6. Advantages and Shortcomings of Decentralized Mechanisms

In comparison with the centralized approaches, decentralized approaches have definite advantages in attacking global problems. New and deeper insights into problems always arise with individuals, and they have a better chance to be creative in smaller groups. I am constantly amazed at the abundance of original ideas which one discovers all over the place if one searches for possible solutions of certain problems, and also about the readiness, the strong motivation and determination of people to tackle them. In perceiving and understanding problems many people are far ahead of their governments and administrations although they may lack detailed knowledge. If these advanced minds should ever leave their niches and try to reach the general public they would generate enormous political pressure. Because decentralized approaches are in general more specific they can be more effective in initiating real changes. It is always astonishing to me how much a single individual can achieve by personal persuasion. It is essential in this context, that not only is a general goal proclaimed, but that a very specific problem is eventually selected for concrete action. Dealing with a problem in the abstract may be enlightening and fruitful for a general orientation, but it can lead to great frustration in the long run, because it keeps us confronted with the problem in its totality, which obviously exceeds our strength and imagination.

The disadvantages of decentralized mechanisms are rather obvious.

- The individual suffers very often from isolation. Confronted with a huge problem in all its full extent and complexity, one is forced to single out a very few particular aspects about which one feels strongly and has particular competence. But one may have trouble in finding people who are willing to participate in tackling a limited issue. If, on the other hand, one does not limit sufficiently one's own scope, one becomes overwhelmed by the immensity of the problem and one's own incapacity to deal with it successfully. It takes, indeed, great powers of conviction and strength not to be driven into frustration but to keep up one's spirits in the absence of any measurable indication of success.
- For private and decentralized organizations it is, in general, more difficult to establish international contacts. This is particularly the case with countries which are politically closed and/or do not have extensive provisions for tourism.
- Private and decentralized organizations have a severe problem in obtaining enough money for their infrastructure. They are rarely able to produce a "product" which could be sold to the public or would attract powerful sponsors.
- This last point applies in particular to the very difficult but important initial phase when a novel approach to a problem is sought. As in fundamental research, this phase is characterized by many trials and only a very few successes. In scientific work we have learned that this basic work, despite its many disappointments, is of decisive importance for all future development and, therefore, government and industry spend a fair amount of money on it. There has so far been no similar investment in the search for solutions to the really vital problems of humanity. Because they could help us to avert great catastrophes the returns on such efforts could be phenomenal.

7. Advantages and Shortcomings of Centralized Mechanisms

There is no doubt that in order to carry out the necessary changes public mechanisms are in the long run necessary. The reasons are:

- tremendous sums of money – probably of the order of trillions of dollars, that is equivalent to the annual world-wide military expenditure – have to be redirected, and governments and central organizations are the only institutions which have the resources to do this;
- governments with appropriate legislation can formulate and set up the necessary framework; they can define new norms and parameters to support and strengthen the required development;

- they can use their ample means of obtaining publicity and their great influence on education to promote strongly the relevant ideas;
- they can invite and motivate other countries to join their efforts by diplomatic missions, international educational programmes, and active cooperation on projects of common interest.

The great disadvantage of public and centralized mechanisms are:

- their intimate connection with the existing power structure, which naturally tries to perpetuate itself by clinging to present conditions advantageous to them, while disregarding the general welfare and the long-range development;
- the tremendous inertia of legislative and administrative processes;
- their inability to establish new norms which optimize future rather than present conditions;
- their concern for the views of a supposed majority of the people, which are commonly projected to be more conservative and inflexible than they really are;
- the normative nature of their methods of implementation which often fall very far short of enforcing what was specifically intended;
- their general tendency to do only the minimum necessary to alleviate the situation rather than going to the roots of the problems.

8. The Right Mix

From the foregoing analysis it is evident that neither centralized nor decentralized mechanisms, neither public nor private mechanisms by themselves will succeed in solving the urgent global problems, but that it will require a certain combination of both. It appears that in the early stages of any initiative the decentralized and private mechanisms will be of decisive importance. All new ideas will come from individuals. It will be in small groups that these ideas will be intensively disputed, criticized and corrected. Only in small groups is there enough motivation and drive to develop them further. The basic handicap to initiatives by small groups will be to get out of their special "groove", that is to go beyond the limits of their particular discipline, culture, and the political and geographical framework from which they started. Our modern societies are highly stratified. To a large extent, they lack institutions where different kinds of people meet to discuss important issues of common concern and where creative friction between very different points of view is generated. Most of us spend nearly all our time in our highly specialized professional circles and the corresponding social environments. "Foreign" concepts intruding into our professional field from outside are watched with great suspicion, and quite often are designated as "sand in the gears" rather than as means of establishing a better interlinkage between different fields and generating fruitful synergies. On the other hand, scientists and members of other professions have proved very effective in crossing geographical borders and political demarkation lines. The active cooperation of scientists all over the world on issues of common interest represents an impressive example of how people in various parts of the world, by direct contact and without involving their respective governments, can join forces to attack common problems. Modern means of communication – direct-dialing telephone, telefax and electronic computer networks – and modern means of information – in particular satellite-transmitted television programmes or simply video tapes on important global issues – may greatly facilitate these horizontal linkages. These means should not be used simply to step up the amount of information, which by its sheer abundance may crush any personal involvement, but they should help in pointing out a variety of very special tasks vital to the general problem. The concentration on very specific and limited issues will enable people of various backgrounds to become personally involved without overstraining their abilities.

Great efforts, therefore, should be made to start a large number of decentralized private groups, each focusing locally on a vital issue linked to global problems, to provide information on the existence and specific goals of such groups such that anybody interested can join, to facilitate communications between the various decentralized groups working on similar topics, and to transfer know-how and advice across all borders – disciplinary, cultural, political and geographical. These groups should not limit themselves to the analysis of the present situation and to general ideas about how to improve it, but from the very beginning they should consider how these ideas could be implemented in some exemplary way, say by initiating and co-ordinating certain pilot projects. Such pilot projects have to be well chosen to obtain a maximum effect with the very limited resources small private organizations can provide. To initiate and organize such pilot projects and to manage their first step special institutions will be necessary. They will need to co-ordinate and integrate the work of the various groups related to the problem area of the project.

It would be very desirable if already at this important initial stage public mechanisms could give substantial support to these groups, just as in all industrialized nations public money is generously given to support basic research; this applies not only to national laboratories but to privately-organized institutions, like the Max-Planck-Society in FRG. I consider it desirable that everybody – man and woman, expert and non-expert, thinker and doer – and not only a self-elected elite of people, should spend a certain percentage of their time reflecting on how the work they are doing can be appropriately modified to satisfy the criteria of a sustainable future; and that governments should spend a reasonable amount of money – at least of the order of the present scientific budgets – to investigate ways to implement viable proposals. This money – like the investment in basic research – will turn out to be very well spent. Besides the immediate positive results in solving the problems in question, it will dramatically increase world-wide cooperation across all borders, and consequently build up mutual confidence and solidarity. The lowered military expenditures will more than compensate for the financial investment necessary for this procedure.

As in scientific-technical developments, where only a small part of the investment goes into basic research and development of a prototype, and the larger part is required to turn out the final product, similarly, to solve the global problems very large-scale investments are necessary. Obviously, this can be handled only by public mechanisms, that is by really big money spent by the various states. Since, hopefully, the feasibility of certain proposals will be explicitly demonstrated by private initiatives, we may be confident that even conservative governments can be eventually persuaded to take the necessary next steps.

9. Examples for Private Decentralized Transnational Activities

There are many examples of private decentralized non-profit organizations involved in transnational cooperation. In connection with the quest for a nuclear test ban and its verification, "Natural Resources Defense Council" (NRDC) in the USA is a particularly impressive example. Their private initiative in setting up seismic monitoring stations in Semipalatinsk in the Soviet Union, and in Nevada in the United States, triggered a whole sequence of events which nearly brought all nuclear testing to a halt. That they eventually did not succeed – some even claim they were not even close – was very unfortunate but this does not detract from their extremely meritorious conduct. And there may be a next time when a US government will prove to be as enlightened as its people.

The peace-oriented movements of scientists, like the "Union of Concerned Scientists" (UCS) and the "Federation of American Scientists" (FAS) in the United States, the Soviet Scientists against Nuclear Threat and for Peace" (SC) in the Soviet Union, the "Scientists" Initiative – Responsibility for Peace' in FRG, and the peace initiative of the medical doctors "International Physicians for the Prevention of Nuclear War" (IPPNW), are other good examples. On the ecological level, "Greenpeace" and the "World Wide Fund for Nature" (WWF) have crossed geographical borders long ago and are now also successfully penetrating the Eastern countries. Of course, Pugwash should be mentioned very prominently in this connection. Pugwash is mainly scientifically-oriented and follows very much the lines of international cooperation in purely scientific fields, like physics, chemistry and biology, although it had some strong political links in its earlier stages. It has developed, therefore, into an important resource organization which diligently and competently analyses relevant global problems, mainly in the security area, and works out and suggests solutions, but, because of its academic posture, the dissemination and implementation of these suggestions are left to other groups to pick up, which unfortunately they rarely do.

It is not the purpose of this paper to describe in detail all these organizations and their transnational work. It is sufficient to state that recent years have seen a large increase in the number of contacts between private groups in East and West, in particular on ecological issues, and in connection with the problem of long-term energy provision. But to my knowledge, only very few of these contacts have led as yet to the establishment of formal international non-governmental organizations. To give concreteness to my theme I intend next to comment on two organizations, the "Global Challenges Network" (GCN) and the newly founded "International Foundation for the Survival and Development of Humanity" (IFSDH) with which I am most familiar. On the GCN and the IFSDH I have written extensively in a recent paper [2]. Both organizations are steadily making their way but are suffering at present somewhat from limited financial resources, in particular in hard currency. The IFSDH, founded by special decree of the Soviet Council of Ministers in October 1984, and publicly announced by General Secretary Mikhail Gorbachev on 15 January 1988 in the Kremlin, is the first international non-governmental organization to be based in the Soviet Union. The GCN was founded one year earlier in FRG to bring together concerned scientists, those involved in grass-roots initiatives, environmental organizations, and government and industry, to face "global challenges". I will not go into any detail on these organizations and their projects but will simply illustrate the type of transnational cooperation which is starting to evolve in connection with a pilot project "Cleaning-up the Baltic Sea" initiated by GCN and Greenpeace in a workshop in July 1987 and financially supported in part by the IFSDH in 1988.

Clearly, the Baltic is one of the most polluted seas posing tremendous problems, aggravated by the fact that sea water is only very slowly exchanged because of the limited connection to the open sea. But this holds similarly for other enclosed seas, such as the Mediterranean. The special situation with the Baltic arises from the fact that it is surrounded by Western, Eastern and neutral countries. Therefore this project proposal has not only an ecological but also a political dimension. It necessarily requires the cooperation of countries with different political and economic systems. The purpose of the Baltic project is to facilitate the development of East-West cooperation through a programme of scientific and technical exchange and skill sharing over several years. Scientists and experts would collaborate on joint investigations, with access to centres of information and advanced communication and analysis facilities. Although much has been done during the past 15 years to stop the degradation of the Baltic, there are still serious environmental problems which must be solved.

To clean up an ocean is, of course, a formidable task and therefore, not quite the type of problem one should select for a beginning. To start an extensive research programme, involving an investigation of the character and the damaging effects of some 50 000 different pollutants, is prohibitively complicated and time-consuming. Any finding will come much too late to prevent the total collapse of the Baltic Sea's ecosystem. To clean it up will, in particular, necessitate the cleansing of the rivers flowing into it. To clean up those rivers, all the sewage and waste disposal dumped into them will have to be filtered and purified. In contrast to cleaning up an ocean, the cleaning up of a single sewage pipe on the territory of a given community is a problem which a community, in principle, should be able to handle. It requires determined citizens and decentralized action to push this project, and a town administration which is willing and able to finance such an undertaking. To support this process there should be an intensive exchange of information on this issue, an extensive transfer of technical know how and generous help to finance effective installations. Since a dead Baltic Sea would hurt all neighbouring states, national boundaries – and in this case even bloc division lines – have no meaning.

In a workshop organized jointly by GCN and the West-German Friedrich-Ebert-Foundation in November 1988, environmental groups of all the states bordering the Baltic, in particular West German BUND (Bund für Umwelt und Naturschutz Deutschland), the Green Movement of Estonia, the Polish Ecological Club, the Swedish-Polish Environmental Society (SPM), and the WWF, decided to start a "Clean Baltic Sea Initiative" with some exemplary measures. They concentrated first on the Polish Vistula river as one of the main polluters, on some large agricultural farms in Schleswig-Holstein and Denmark, as well as in Soviet Estonia and the Ladoga lake area. They sought to monitor the phosphate and nitrate input leading to the harmful euthrophication, and to review the difficult questions of financing and optimizing financial investments by looking into methods such as Debt/-Nature Swaps, Debt Relief, Technology Funds, Risk Capital Investments and Clean Baltic Fund. For the project it will be vitally important to involve and mobilize not only the administrative authorities but also ordinary citizens. This requires better communication on these issues and offers extended possibilities for personal engagement and cooperation. To support this, an additional project: "A measurement and communication network for environmental control and monitoring accessible to the general public", was proposed and started by GCN, and is at present in the test phase in FRG.

It is my impression that all these direct contacts between private groups in East and West on issues of common interest will prove extremely effective and will speed up solutions of these very difficult and long-term problems. In addition, they will constructively improve the political climate between East and West. I hope that similar steps can be taken to bridge the gap between North and South – a task which will prove to be even more difficult but nevertheless should be taken up soon and with conviction. There is no doubt in my mind that any progress along the lines outlined in this paper will also contribute to overcoming the present intolerable North-South imbalance.

References

1 In *Global Problems and Common Security* (J. Rotblat and V. I. Goldanskii, eds), Springer, Heidelberg, 1989 p. 259

2 *New Paths to Global Cooperation: The Global Challenge Network and the International Foundation for the Survival and Development of Humanity*, paper presented to the 50th Pugwash Symposium Prague, April 1988. (Pugwash Newsletter, May 1988 p. 145)

Appendices

Statement of the Pugwash Council on the 39th Pugwash Conference, Cambridge, Massachusetts, 23–28 July 1989 Building Gobal Security Through Cooperation

The 39th Pugwash Conference on Science and World Affairs met in Cambridge, Massachusetts, from 23 to 28 July 1989 on the theme of "Building Global Security Through Cooperation". The meeting was attended by 177 scientists, other scholars, and public figures designated as regular participants, 19 members of Student/Young Pugwash groups from around the world, plus some two dozen additional invited speakers, observers from national and international organizations, and guests.

The attendees came from 42 countries and included 4 Nobel Laureates, 3 members of the Soviet parliament, 7 current and former ambassadors, 4 active and retired generals, and numerous past and present heads of universities, research institutes, and government agencies.

The Conference, which took place at the Marriott Cambridge Center hotel with special sessions at the American Academy of Arts and Sciences and MIT's Kresge Auditorium, was hosted by the US Pugwash Committee, which is affiliated with the Committee on International Security Studies in the American Academy. Major sponsors for the meeting included the MacArthur, Sloan, General Service, Levinson, and Mertz-Gilmore Foundations, the New Horizons Trust, and MIT.

Speeches of Welcome at the opening session of the Conference were given by Massachusetts Governor Michael Dukakis and Senator John Kerry (D-Mass). The keynote speaker at the Conference banquet was Senator Timothy Wirth (D-Colo), who stressed the interaction of problems of environment, development, and security and the need for greatly expanded cooperative international initiatives to cope with the array of threats confronting humankind.

President Bush sent a letter of greetings to the participants, noting that "By fostering the international exchange of ideas and opinions on the key science and public policy issues of the day, Pugwash has made a valuable contribution to world peace and understanding". Letters of greetings were also received from the President of Hungary, the Presidents of the Soviet and Hungarian Academies of Sciences, and a number of international organizations.

The atmosphere that permeated the discussions at the Conference was, as has often been the case in Pugwash meetings, a combination of hope and alarm. The sense of hope – stronger and more widespread than at any time within memory – stems from the *unprecedented opportunities* that now seem within reach for cooperative action against the greatest worldwide threats to security: armed conflict, weapons buildups, economic inefficiency and stagnation, hunger, under-development, spiralling debt, environmental degradation, abuses of human rights. The sense of alarm comes from the conviction – also very widespread among the participants – that *if the opportunities are not seized and vigorously exploited during the next decade, it will be too late*: the problems will have grown by then beyond the possibility of solution.

Eight focuses of the needed action were the topics of the Working Groups that met in repeated parallel sessions throughout the week-long Conference: (1) strategic forces and their limitation; (2) European security; (3) chemical weapons arms control; (4) nuclear weapons proliferation and testing; (5) the arms trade and regional conflict; (6) the role of the UN and other multilateral organizations in international security; (7) fuels, forests, and carbon dioxide – the energy/environment/development interaction; and (8) agendas for East-West and North-South cooperation. Plenary sessions treated "Recent Developments in Arms Control and Other Aspects of Science and World Affairs", "Current Issues in Arms-Control Verification", and "Future Priorities for the Pugwash Conferences", among other topics.

The present statement on the issues addressed at the Conference draws on the working-group and plenary-session discussions, but its formulation is the responsibility of the Pugwash Council alone. We have not attempted to be comprehensive – many important points discussed at the Conference are not mentioned here – and our statement has not been reviewed or endorsed by the other attendees. Participants in Pugwash Conferences take part as individuals, not as representatives of their governments or institutions, and on some issues they hold a wide range off views. Summaries of the discussions in the Working Groups have been prepared by the Rapporteurs in consultation with the Convenors of these Groups and will become part of the published Proceedings of the Conference, along with the texts of the major speeches from the plenary sessions.

Clearly, the single most important contributing factor in the new possibilities for building global security through cooperation is the remarkable transformation that has been underway in the Soviet Union under the names of *Perestroika* and *Glasnost*. The "new political thinking " that has initiated and sustained this transformation bears a striking resemblance to the "new thinking" embodied in and called for in the Russell-Einstein Manifesto of 1955 – the thinking that led to the organization of the Pugwash Conferences in 1957 and that has been pursued through these meetings ever since.

In any case, the application of these ideas in the Soviet Union under Mr Gorbachev has led to tremendously promising changes in the internal politics as well as in the foreign policies of that country; and these changes, and the reactions to them in other countries, have made possible a rate of progress in arms control, international tension reduction, and security building that would have been almost unimaginable only a few years ago.

At the same time, the nature and pace of these radical changes inevitably has caused resistance from conservatives within the Soviet Union, as well as revealing more starkly than before the magnitude of the economic predicament in which that country finds itself. If the process of *Perestroika* is derailed by the frictions accompanying it (including discontent with problems *perestroika* and *glasnost* have made clearer but did not cause), or by inappropriate or inadequate responses from the West, the result will be a catastrophic setback not only for the Soviet Union but also for the rest of the world. The new opportunities for cooperative approaches to building global security could disappear even more rapidly than they have materialized.

The rate of growth of the dangers and problems for which such cooperative approaches are the only solution – not only in the military sphere but also in the economic and environmental ones – means that the world will not get another chance. Leaders and publics in the West, in particular, should not allow residues of a Cold War mentality, or simply excessive caution, to impede the build-down of the East-West military confrontation or the build-up of the East-West economic cooperation. On the success of these initiatives depends not only the success of perestroika for the Soviet Union, but also the prospects for building more secure and prosperous societies world-wide.

In the field of strategic arms control, new thinking on both sides has been making inroads although there is still far to go. A START agreement along the lines now envisaged in the official negotiations, representing about a 50 per cent cut in strategic forces by the counting rules being considered, would not in itself make the world a great deal safer; nor would it necessarily save money; but nonetheless should be achieved as quickly as possible because it moves us in the only sensible direction – towards still deeper reductions.

There are no insurmountable verification obstacles to an early START agreement. Some kinds of strategic forces are more difficult to verify than others, but there are no problems that cannot be overcome if both sides really want an agreement. Similarly, there are no good reasons for allowing the ABM issue to block a START agreement: no promising missile defenses are in sight that could provide a plausible excuse to violate the ABM Treaty; the only reliable way to reduce the number of nuclear weapons that could strike one's country is to negotiate a reduction in the number of those weapons possessed by the other side.

Concerning the appropriate goal of strategic arms reductions beyond START, it would be worthwhile to aim at an intermediate level in the range of 10 to 20 times fewer strategic warheads than exist today, with careful attention to minimizing vulnerability of the resulting "minimum deterrent" force to a first strike. Although there is less agreement about the feasibility of achieving a world without any nuclear weapons whatever, this remains the goal of Pugwash and we continue to seek prescriptions for bringing a zero-nuclear-weapons world within reach. Elimination of all nuclear weapons stationed on European territory would be a good start, and may become practical soon.

Continued maintenance of nuclear arsenals of any size in any country, and continued adherence to the idea that any country needs or is entitled to a nuclear deterrent, seems to us to be incompatible in the long run both with non-proliferation and with putting an end to technological arms races. If some countries insist on keeping nuclear weapons, others will want them too, even if logic indicates that possession of nuclear weapons decreases a country's security. And it is difficult to imagine a situation in which one or more countries retain a class of weapons indefinitely without trying to "modernize" them – the stuff of arms races.

The 1968 Non-Proliferation Treaty (NPT) has not been signed by some countries, and is under increasing criticism by some countries that have signed it. Its critics point to the inherently discriminatory character of the NPT, and to the failure of the nuclear-weapon-states to comply with their obligations under Article VI to negotiate in good faith towards nuclear disarmament. There is disagreement in the international community between some countries that espouse the goal of replacing the NPT with a security regime in which all countries participate on an equal footing, and other countries that stress the need to preserve the NPT in spite of its imperfections.

It is highly desirable, in any case, that the following measures to strengthen the non-proliferation regime and to discourage the modernization of existing nuclear arsenals be undertaken as soon as possible and in any case before 1995: (1) the achievement of a Comprehensive Test Ban on nuclear explosions, which could most expeditiously be accomplished through a conference to amend the Partial Test Ban Treaty of 1963 (a process that has already begun); (2) a global ban on the production of weapons-grade fissile material; (3) an agreement to transfer the fissile material from retired nuclear weapons to peaceful uses under IAEA safeguards.

Concerning chemical weapons (CW), we believe that the total, global, verifiable prohibition of chemical weapons that has been under serious negotiation in Geneva since 1984 could and should be completed within two years. This crucial goal has been brought closer by agreement in principle on the key issue of challenge inspections, by growing concerns about chemical-weapons proliferation, and by the public commitments of President Bush and President Gorbachev to conclude the treaty.

Some details of verification procedures for the CW Treaty remain to be worked out, but there should be no insurmountable problems. It must only be remembered that what is sought is a balance between verification too loose to provide needed assurance of compliance and verification too tight to be compatible with legitimate chemical-industry operations; the aim is for all states to conclude they are better off inside the treaty than outside it.

Concerning conventional armed forces in Europe, the new prospects for deep cuts and restructuring to reduce the potential for offensive operations are dramatic and extremely encouraging. We are pleased at the contribution that Pugwash Workshops held since 1984 on the approach and detailed prescriptions for such reductions have made towards bringing about these possibilities. It is important now to maintain the momentum towards early agreement, and not to lose sight of the primary goals of increasing stability and defensivity in the thicket of negotiations over numbers.

The role of non-military factors in European security is increasing and is making the solution of the military questions much easier. Important above all in this connection is continued work to strengthen economic cooperation between Eastern and Western Europe, including especially measures aimed at strengthening the economic underpinnings of pluralism and openness in Eastern Europe and the Soviet Union.

Armed conflict, a declining threat in the North, is a continuing reality in the South. The human and economic toll that has been taken by such conflicts in the South since the end of the Second World War is staggering. Although the root causes of many of those conflicts have been and continue to be found in local conditions in the South, the North has often played important roles through policies that helped create or maintain these conditions, through ill-advised interventions, and through transfers to weapons that have increased the intensity or duration of conflicts in the South.

The burden of resolving local conflicts must reside above all with the local powers, but we must also expect more from the countries of the North in the form of assistance in conflict settlement, cooperation in the implementation of nuclear-weapon-free zones, restraint in the transfer of weaponry and weapons technology, and a positive example through faster progress towards arms reductions in the North. Future arms-control agreements in the North should contain provisions that weapons affected by these agreements be destroyed or placed in secure storage, rather than being transferred to other countries.

The role of the United Nations in building global security can and should be strengthened in the areas of regional dispute resolution, facilitation of the reduction of arsenals of mass-destruction weapons, and peacekeeping. Permanent peacekeeping forces ought to be established under the auspices of the Security Council, with adequate financial and logistical support. At the same time, the need to reconsider the structure of the Security Council itself should be apparent from the circumstance that, at present, all of the permanent members are nuclear-weapon-states; the impression that this is a criterion for admission to this select circle is unfortunate and dangerous.

The United Nations may also have a role to play in safeguarding the peaceful uses of space – including the functions of satellites for communication, earth-resources assessment, and scientific studies – and preventing its weaponization. A separate UN Outer Space Agency might be the appropriate vehicle for these purposes.

Large-scale environmental problems, particularly the risks posed by global climatic changes induced by "greenhouse" gases, have the potential to cause widespread increases in human misery as well as to aggravate other forms of tensions. The mechanism by which this occurs may include reduced availability of food, water, and fibre, altered patterns of disease, increased disparities between haves and have-nots, creation of new groups of environmental refugees, and disputes over the responsibility for and costs of environmental protection.

The production and use of energy resources in both developed and developing countries are most often pursued with insufficient regard to the potentially enormous long-term environmental costs. In developing countries, the current level of energy use must be increased to provide for basic human needs, but large improvements in energy efficiency are possible everywhere and must be a key ingredient in any sensible strategy to reduce greenhouse-gas emissions.

The implementation of efficiency improvements, the development of environmentally less disruptive energy technologies suited to regional conditions, and the development of improved monitoring and analysis of large-scale environmental problems are all suitable focuses for greatly expanded scientific and technological cooperation. Much of this work could make use of human and material resources liberated from the military sector.

The transfer of resources from the poor countries in the South to the rich countries in the North in order to pay back massive debts is hindering economic development in the South, encouraging tremendous environmental destruction, and worsening North-South tensions. The debt problem can only be resolved through a cooperative approach including, for example, more equitable terms of trade, the issuance of Special Drawing Rights to Third World countries, establishment of a special fund to be financed by royalties and taxation of activities in the global commons (i.e., the oceans, space), and transfers of funds out of military expenditures and into development programmes.

The issue of effective use of science and technology for development was intensively analysed in the 1970s, including important work done in Pugwash on a code of conduct for technology transfer. Since the formulation of the Vienna Programme of Action and its approval by UN member states in 1979, however, efforts in this direction have languished. Among other organizations that may be expected to play a part in the revival of such efforts is the new International Foundation for the Survival and Development of Humanity.

As we stressed in the Dagomys Declaration following last year's Pugwash Conference in the Soviet Union, the threats to global security are not only military and political but also economic and environmental; and the problems and fates of East and West are inextricably entwined with those of the South. Shaping a prosperous, secure, and sustainable set of societies worldwide will require a comprehensive and cooperative approach to this complex of problems. The knowledge needed to do so is largely in place. It is time for scientists, politicians, and publics to work together to muster the political will.

Statement of the Pugwash Executive Committee on the Fourth NPT Review Conference in 1990

1. The most important relevant events since the Third Review Conference in 1985 have been:
   (a) the bilateral Intermediated Nuclear Forces (INF) Treaty between the Soviet Union and the United States to eliminate whole classes of intermediate-range and shorter-range nuclear-armed missiles under an extensive verification regime;
   (b) the agreement in principle to seek a treaty (now under negotiation) aimed at approximately 50 per cent reductions in the strategic nuclear forces of the two countries; and
   (c) the declaration of General Secretary Gorbachev and President Reagan, at the summit meetings preceding these agreements, that they share the goal of a nuclear-weapon-free world.
2. These events can reasonably be described as an overdue but very welcome beginning by the nuclear-weapon states towards fulfillment of their obligations under Article VI of the NPT. The prior persistent failure of these states to demonstrate a serious commitment to fulfillment of those obligations has been the greatest threat to the stability of the NPT and one of the principal reasons that some states have not joined it.
3. Other favourable indications since 1985 for the future of the NPT include: the increase in the number of signatories to 139; the absence of any overt additions to the roster of nuclear-weapon states; the declared interest of South Africa in signing the NPT; and the coming into force of the Rarotonga Treaty, creating a nuclear-weapon-free zone in the South Pacific that covers one sixth of the planet and is contiguous to the nuclear-weapon-free zone created by the Tlatelolco Treaty and the demilitarized zone created by the Antarctic Treaty.
4. Certain unfavourable developments since 1985, on the other hand, have reinforced concerns about the stability of the non-proliferation regime. These include:
   (a) disclosures about the atomic energy programme of Israel by a former employee, indicating a greater nuclear-weapon capability in that country than had been supposed previously and thereby aggravating threat perceptions in the region;
   (b) persistent rumours that South Africa and possibly the near-nuclear states in Asia and Latin America have been progressing in the development of nuclear weapons;
   (c) reported assistance to the weapons programmes of these countries by transfers of weapons-relevant materials and technology from other States;

(d) the spread of ballistic-missile technologies to additional States; and
(e) continuing "vertical" proliferation in the form of development and deployment of new nuclear weapons systems by the major powers.

5. Probably the most unfavourable factor of all is the continued adherence of nuclear-weapon states to explicit or implicit doctrines that insist on the need for maintaining national nuclear forces as deterrents against nuclear or even conventional attack. It needs to be recognized that one of the gravest liabilities of nuclear-deterrence strategy, overt or covert, is its fundamental incompatibility with non-proliferation goals, inasmuch as there is no logical basis for denying the "right" to a nuclear deterrent to some States while according it to others.
6. To move ahead towards a more robust and more equitable non-proliferation regime requires decisive new actions to exploit the favourable factors mentioned above while minimizing the effects of the favourable ones. The most important elements of the needed programme are:
   (a) *Adoption of a comprehensive action plan, with a specific timetable, for stopping and reversing the nuclear arms race* among the major nuclear-weapons powers, including (i) strict adherence to the spirit as well as the letter of the INF Treaty, (ii) early completion of an agreement for cuts of 50 per cent or more in the strategic nuclear arsenals of the United States and the Soviet Union, as a prelude to further reductions, (iii) early agreement on a prescription for including the other nuclear-weapon states in the build-down of nuclear arsenals, (iv) an immediate moratorium on all nuclear testing, followed by negotiation of a comprehensive test ban treaty, (v) a universal production cutoff on weapons-grade fissile material, under nondiscriminatory international control; and (vi) rapid progress towards bringing under control the other dimensions of military competition and confrontation that are intertwined with and aggravate the nuclear dimension (e.g., chemical weapons, conventional forces, and the weaponization of space).
   (b) In conjunction with such progress, and in any case as soon as possible, *achievement of a formal commitment by all nuclear-weapon states not to be the first to use nuclear weapons* (as already undertaken by China and the Soviet Union), not to structure their forces in ways that confer the appearance of first-use intention or capability, and not to threaten the use of nuclear weapons.
   (c) *Steps to strengthen the Non-Proliferation Treaty, in preparation for its extension in 1995*, beginning with immediate initiation of a process, involving non-signatory states as well as signatories, aimed at agreement on means for enhancing the stability and equity of the non-proliferation regime, including (i) conferral or increased authority and resources to the International Atomic Energy Agency for safeguards implementation, (ii) an end to all overt and covert assistance by states to the nuclear-

weapons programmes of other states, and (iii) inducement of current non-signatories to sign the NPT.

(d) *Universal acknowledgement that the complete elimination of nuclear weapons, and general and complete disarmament, are essential ultimate goals* towards which the world community can and must undertake steady progress.

Acronyms

ABM	anti-ballistic missile
ACDA	Arms Control and Disarmament Agency
ALCM	air-launched cruise missile
ALPS	Accidental Launch Protection System
APC	armoured personnel carrier
ASAT	anti-satellite weapon
ASEAN	Association of South-East Asian Nations
ATTU	Atlantic to the Urals
AWACS	airborne warning and control system
BP	brilliant pebbles
C3I	command, control, communications and intelligence
CD	Conference on Disarmament
CDC	command-destruct capability
CFE	conventional forces in Europe
CSBM	confidence - and security-building measures
CTA	compensatory trade agreements
CTB	comprehensive test ban
CW	chemical warfare
CWC	chemical weapons convention
DEFCON	defence condition
ECA	Economic Commission for Africa
ECM	electronic counter-measures
ERIS	Extra-atmospheric Intercept system
G7	Group of Seven (industrialized countries: Canada, France, FRG, Italy, Japan, UK, USA)
GATT	General Agreement on Tariffs and Trade
GCD	general and complete disarmament
GDP	gross domestic product
GNP	gross national product
HRGC	high resolution gas chromatography
HRLC	high resolution liquid chromatography
IAEA	International Atomic Energy Agency
IBRD	International Bank for Reconstruction and Development
ICBM	intercontinental ballistic missile
IMF	International Monetary Fund
INF	intermediate nuclear forces

IPCC	Intergovernment Panel on Climate Change
IRIS	International Research Institute for Seismology
JVE	Joint Verification Experiment
LDC	less developed country
MBFR	mutual and balanced force reduction
MILSTAR	high-altitude communications satellites
MIRV	multiple independently targetable re-entry vehicle
NATO	North Atlantic Treaty Organization
NCA	National Command Authority
NIEO	New International Economic Order
NMR	nuclear magnetic resonance
NNPA	Nuclear Non-Proliferation Act
NORAD	North American Air Defense Command
NPG	(NATO's) Nuclear Planning Group
NPT	Non-Proliferation Treaty
NRDC	Natural Resources Defense Council
NWFZ	nuclear-weapon-free zone
OAS	Organization of American States
OAU	Organization of African Unity
OECD	Organization for Economic Cooperation and Development
OPEC	Organization of Petroleum Exporting Countries
PAL	permissive action link
PTBT	Partial Test Ban Treaty
R&D	research and development
SAC	Strategic Air Command
SACEUR	Supreme Allied Commander Europe
SAM	surface-to-air missile
SBI	space-based interceptor
SDIO	Strategic Defense Initiative Office
SIPRI	Stockholm International Peace Research Institute
SLBM	submarine-launched ballistic missile
SLCM	sea-launched cruise missile
SNF	short-range nuclear forces
START	Strategic Arms Reduction Talks
TTBT	Threshold Test Ban Treaty
UNCTAD	United Nations Conference on Trade and Development
UNEP	United Nations Environment Programme
WTO	Warsaw Treaty Organization

Notes on the Contributors

Muthiah Alagappa (Malaysia) international relations. Institute of Strategic and International Relations, Kuala Lumpur

Abdel Monem Said Aly (Egypt) international relations. Senior researcher and Head of the International Relations Research Unit, Al-Ahram Center for Political and Strategic Studies, Cairo

Kirill Babievsky (USSR) organic chemistry. Professor of Organic Chemistry at the Institute of Organoelement Compounds, Moscow

Frank Blackaby (UK) disarmament and arms control. Former Director, Stockholm International Peace Research Institute

Julio Carasales (Argentina) international relations - disarm- ament. Ambassador in the Argentine Foreign Ministry. Professor of Argentine Foreign Policy, University of Belgrano, Buenos Aires

Jonathan Dean (USA) arms control. Arms control adviser. Union of Concerned Scientists, Washington D.C. Former US Ambassador to Committee on arms control negotiations

Saadet Deger (Sweden) economics. Head of military expenditure project, Stockholm International Peace Research Institute

Hans-Peter Dürr (FRG) elementary particle physics. Director of Werner-Heisenberg Institute for Physics, Max Planck Institute for Physics and Astrophysics, Munich

Omran El-Shafei (Egypt) international relations. Retired member of UN Advisory Board for Disarmament Matters. Member of Human Rights Committee (International Covenant on Civil and Political Rights). Former Assistant Foreign Minister

Nikolai Enikolopov (USSR) chemical physics. Vice Chairman, Soviet Peace Fund and Director of Institute of Synthetic Polymer Materials, USSR Academy of Sciences, Moscow

Esmat Ezz (Egypt) chemical and nuclear defence. Retired Major- General, former Director of Scientific Research Branch of the Egyptian Armed Forces, Cairo

Essam Galal (Egypt) medical practice and pharmacolgy. Adviser to the Academy of Defence Research and Technology, Cairo

Virginia Gamba (Argentina) strategic studies. Visiting Professor, Department of War Studies, King's College, London

Richard Garwin (USA) physics. IBM Fellow and Science Advisor to the Director of Research; Adjunct Professor of Physics, Columbia University, New York

Alexander George (USA) political sciences. Professor of International Relations at Stamford University, California

Jozef Goldblat (Switzerland) international relations. Lecturer and Senior Research Fellow, Programme for Strategic and Inter- national Security Studies, Graduate Institute of International Studies, Geneva

José Goldemberg (Brazil), physics. President, University of Sau Paulo. Minister for Education

Kurt Gottfried (USA) physics. Professor of Physics, Newman Laboratory of Nuclear Studies, Cornell University, New York

Klaus Gottstein (FRG) science and public policy. Director, Research Unit on Science and Society, Max Planck Institute, Munich

Anatoly Gromyko (USSR) social sciences. Director of African Institute, USSR Academy of Sciences, Moscow

He Gulian (China) mathematics. Professor, Institute of Applied Physics and Computational Mathematics, Beijing

John Holdren (USA) physics. Professor of Energy and Resources, University of California. Chairman of the Pugwash Executive Committee

Enrico Jacchia (Italy) strategic studies. Director, Centre of Strategic Studies. Former Director, Nuclear Security Control, Commission of European Communities, Rome

Zhu Jiaheng (China), physics. Professor, University of Beijing

Karlheinz Lohs (GDR) chemistry/toxicology. Director, Research Department on Chemical Toxicology, Academy of Sciences, Leipzig

Sven Lundin (Sweden) chemical and biological weapons developments and disarmament. Head, CBW Programme at Stockholm International Peace Research Institute

Irving Mintzer (USA) environment and energy. World Resources Institute, Washington DC

Milo Nordyke (USA) geophysics. Head, Center for Technical Studies of Security, Energy and Arms Control, Lawrence Livermore National Laboratories

Surendra Patel (Finland) development economics. Senior Advisor, World Institute of Development Economics Research

Joseph Rotblat (UK) physics. Emeritus Professor of Physics at the University of London. President of the Pugwash Conferences

Lawrence Scheinman (USA) political science and international relations. Professor of International Law and Government; Acting Director, Peace Studies Program, Cornell University, New York

Jane Sharp (Sweden) political science. Senior Researcher, Stockholm International Peace Research Institute

Jasjit Singh (India) arms control. Director, Institute for Defence Studies and Analyses, New Delhi

Bhalchandra Udgaonkar (India) theoretical physics. Senior Professor at Tata Institute of Fundamental Research, Bombay. Former Deputy Chairman of Planning Commission

Laszlo Valki (Hungary) international law and international relations. Professor of International Law at Eötvös University, Budapest. Member of Advisory Board of Hungarian Government

Subject Index

GPSR Compliance

The European Union's (EU) General Product Safety Regulation (GPSR) is a set of rules that requires consumer products to be safe and our obligations to ensure this.

If you have any concerns about our products, you can contact us on ProductSafety@springernature.com

In case Publisher is established outside the EU, the EU authorized representative is:

Springer Nature Customer Service Center GmbH
Europaplatz 3
69115 Heidelberg, Germany

Batch number: 10844669

Printed by Printforce, the Netherlands